The Coal Question

The Coal Question

Political economy and industrial change from the nineteenth century to the present day

Ben Fine

Routledge
London and New York

First published 1990
by Routledge
11 New Fetter Lane, London EC4P 4EE

Simultaneously published in the USA and Canada
by Routledge
a division of Routledge, Chapman and Hall, Inc.
29 West 35th Street, New York, NY 10001

© 1990 Ben Fine

Phototypeset in 10pt Times by
Mews Photosetting, Beckenham, Kent
Printed and bound in Great Britain by
Biddles Ltd, Guildford and King's Lynn

All rights reserved. No part of this book may be
reprinted or reproduced or utilized in any form or
by any electronic, mechanical, or other means, now
known or hereafter invented, including photocopying
and recording, or in any information storage or
retrieval system, without permission in writing from
the publishers.

British Library Cataloguing in Publication Data

Fine, Ben
    The coal question: political economy and industrial
change from the nineteenth century to the present.
    1. Great Britain. Coal industries. History
    I. Title
    338.2'724'0941
    ISBN 0-415-04384-0

Library of Congress Cataloging in Publication Data

Fine, Ben.
    The coal question : political economy and industrial change from
the nineteenth century to the present day / Ben Fine.
    p.   cm.
    Bibliography: p.
    Includes index.
    ISBN 0-415-04384-0
    1. Coal trade – Great Britain – History. 2. Coal leases – Great
Britain – History. I. Title.
HD9551.5.F49   1990
338.2'724'0941 – dc20   89-10346
                                          CIP

For Pheroze Cawasji Engineer

Contents

Acknowledgements ix
Introduction xi

Part one Monopoly and coal

1 Monopoly capitalism and the coal vend 3
2 Cartels and rationalization in the 1930s 19

Part two Coal royalties

3 Royalty or rent: what's in a name? 35
4 Royalties: from private obstacle to public burden? 51

Part three Cliometrics and coal

5 Returning to factor returns: the late nineteenth century coal industry 71
6 Returns to scale in the interwar coal industry 83
7 The diffusion of mechanical coal cutting 95

Part four Towards privatization?

8 The commanding heights of public corporation economics 111

Contents

9 **Privatization and property rights: from electricity to coal**   138

10 **Coal: the ultimate privatization**   173

   Notes   187
   References   201
   Index   218

Acknowledgements

Much of the research for this book, which spans a fifteen-year period, has been done in collaboration with others. I am deeply indebted to those colleagues involved, and they have often been co-authors. I have benefited from research grants from the Leverhulme Trust, the ESRC, the Nuffield Foundation and the Central Research Fund of the University of London. Library facilities and assistance, too numerous to list, have been invaluable. Many others have helped me in a variety of ways. To all concerned, my most sincere thanks.

Introduction

The coal industry has occupied a central position in the British economy over the last two centuries. Although its employment (over a million workers) and output (almost 300 million tonnes of which a third were exports) peaked around the First World War, its rise to such heights and its subsequent decline have been dominant features of the British industrial landscape. Moreover, the conflicts that have surrounded the history of coal, most notably the strikes of 1926 and of 1984/5, have had a more general impact on British society.

Not surprisingly, then, the changing fortunes of the industry have been examined by an economic history that has responded according to the different problems that the industry appears to present for analysis at different times during its development. This book is primarily about the economic analysis that has been employed to understand the coal industry. As such, it is highly critical but it is not and cannot be simply a volume devoted to the history of economic thought. Rather it also depends upon its own analysis of some of the problems that have plagued the British coal industry. For otherwise, it would not be possible to assess the strengths and weaknesses of existing economic histories on the basis either of the theory that they employ or of the empirical developments that they choose to focus upon (or to neglect).

Part I is concerned with the theory of monopoly. From the sixteenth century until the mid-nineteenth century, when the railways opened up new inland sources of supply, the London market for coal was dominated by shipments from the North East, the proverbial coals from Newcastle. In an early and neglected work – his PhD thesis – Paul Sweezy, later to become the United States' most prominent Marxist economist, analysed the coal trade in terms of the attempts to create and hold together a monopoly in the form of a cartel. The close connection is brought out between this analysis and the one that Sweezy was to develop for the modern era of 'monopoly capitalism'. A correspondence is revealed between the theory used to explain the early coal industry and the theory used to characterize the most developed stages of capitalism. This

Introduction

suggests the conflation of monopoly of a market around the turn of the Industrial Revolution with monopoly as a generalized feature of contemporary capitalism. Thus, the (stagnationist) monopoly capital school is seen to have derived its analytical roots from an understanding inappropriately drawn from earlier periods of capitalism. This is liable to lead to errors in analysis for each of the objects of study.

During the 1930s, the British coal industry enjoyed the benefits of a state-organized cartel. In a standard interpretation of this, Kirby has argued that the result, as predicted by monopoly theory, was to restrict output, hold up prices and to impede reorganization of an industry cushioned by monopoly profits. On both empirical and theoretical grounds, it is shown that this interpretation is incorrect and that cartels as such do not impede reorganization, since the latter can boost profitability from cost reduction, irrespective of the gains to be made by market restrictions.

Part II revolves around the theory of rent. As mining depends upon the fertility of reserves, it is hardly surprising that rent theory should be prominent in some analyses of the coal industry. Indeed, the Ricardian notion of diminishing returns to land, as developed for agriculture but applied to the coal industry, has dominated and continues to dominate thinking about the long-term future of the industry. As the best coal reserves are extracted so mining must move on to increasingly poor working conditions, raising costs and reducing the economic attraction of coal.

This particular point of view emerged most prominently a hundred years ago in a debate over whether coal royalties were comparable to an economic rent or not. This debate is reviewed and it is shown that it was a product of its time in so far as the conditions of access to the coal-bearing lands were increasingly proving a problem for British mine owners. But the debate over royalties failed to address this issue since it was more concerned, in theory, with physical, geological conditions rather than being concerned, in practice, with the pattern of land ownership and conditions of rights of access. Paradoxically, the miners' own approach to the problem, as one of practical measures such as nationalization of the royalties, was unversed in economic theory and, consequently, more able to recognize the issues involved. Through both theoretical and empirical study, the chapters in this part reveal the importance for rent theory of examining the specific and changing conditions under which capital has access to the land.

Part III deals in coal and cliometrics. The latter tends to divide economic historians into those that love it and those that hate it. There are good reasons for this. In general, but not of necessity, the cliometrics of individual industries employs the abstract and unrealistic propositions of neoclassical economics. In this, analysis of an industry is reduced

Introduction

to the measurement of factor inputs and factor rewards. As it were, money measures the value (in a causal sense) of everything. The economic historian trained in sophisticated accounts of cause and effect can only be offended by such methods. Such an outlook is supported here by critically reviewing one of the early and classic studies within cliometrics, that of the late nineteenth-century coal industry by McCloskey.

But cliometrics is also to be welcomed because of its commitment to making full use of the available data. Sophisticated accounts of cause and effect are all very well but they need to be empirically substantiated. It is shown, contrary to the standard interpretation associated with Buxton, that there were increasing returns to scale in the British coal industry in the interwar period. Consequently, the failure to amalgamate in that period can be thought of as an entrepreneurial failure. On the other hand, it is also shown that such mechanization as did take place in the mines during that time, and it lagged behind competitors, was geared towards best practice technology. Accordingly, entrepreneurs appear both to have failed and to have succeeded. This paradox is resolved by removing the standard debate over entrepreneurial failure away from the role of the atomized behaviour of individual capitalists and situating it, instead, on the terrain of collective action within the developing economic structures.

Part IV confronts contemporary problems of the coal industry. These have been prominent due to the miners' strike of 1984/5 and the effects of the Tory Government's privatization programme. The plans for electricity have profound implications for coal and, in part, pave the way for its own, the 'ultimate', privatization.

The analysis here questions much of the conventional wisdom concerning the nationalized industries, the privatization programme and the 'uneconomics' of the coal industry. Whilst the neo-Austrian justification for privatization through a general commitment to *laissez faire* is readily dismissed, along with the Government's own rationale, together they reveal the extent to which privatization represents a continuity, not so much a break, with the economic policies of the past.

More challenging analytically has been the birth of a new orthodoxy in public corporation economics. This stresses the importance of regulation and competition to deal with natural and artificial sources of market imperfections, and it is predominantly concerned with static problems of efficiency at the expense of the dynamic problems of economic growth and the failure of the British economy in this respect. As such, it is shown to be both short-sighted and analytically limited in economic, social and historical dimensions.

This analysis is supported by specific discussion of the privatization of both electricity and coal. In addition, the complexity of the issue of ownership is demonstrated by reference to the miners' strike of 1984/5, which can be seen as an intense conflict over the nature and balance of property rights – rather than these being seen as a residual claim to a stream of income.

It should be apparent that this book is extremely ambitious in its coverage, dealing with the history of an industry over two centuries and the accompanying economic and historical thought. No claims are made for anything approaching comprehensive coverage. Indeed, many of the main aspects of the subject matter are covered only cursorily or even not at all. This is partly because there are many standard texts, not least for the economic history of the industry, especially with the publication of the 'official' histories. It is also because the intention has been more to bring to the fore alternative interpretations and causative factors in economics and economic history. This, more than coal, is what holds the book together and, hopefully, it will prove its worth to readers other than those simply interested in the particular topics covered.

In the study of the contemporary situation, economic change and its associated literature are unfolding as the book is being written. As a rough and ready measure, this study covers the period up to the end of 1988. Although some important contributions to the literature may have been overlooked, equally a few references have slipped through the turn-of-the-year net in the opposite direction – by way of compensation.

Part one
Monopoly and coal

Chapter one

Monopoly capitalism and the coal vend

Paul Sweezy has been a central figure in the understanding and development of Marxist political economy both in the United States and more widely in the West.[1] For a long period, much like his counterparts in the UK, Maurice Dobb and Ronald Meek, his was almost a lone voice along with his close collaborators, most notably Paul Baran. Much has changed in the last twenty years with the renewed intellectual interest in Marxism following the student activism of the 1960s – so much so that Marxism has even attained the status of academic respectability. This has meant that, whilst there was always opposition to the Sweezy problematic (as represented by the 'Monopoly Capital' or Monthly Review school), only in recent years has it been substantially criticized and counterbalanced by alternative schools of Marxism.

In particular, there has been a lessening sympathy for underconsumptionist theory; for the notion that monopoly and competition are inversely related; for the validity of the concept of the potential surplus and its compatibility with value theory; and, as a more general aspect of the latter, that monopoly and competition and accumulation can be analysed independently of the labour (value producing) process, whatever the merits of Braverman's (1974) seminal contribution. In addition, unfortunately on the margins of political economy, rather than its occupying a central position of debate as within the often supposedly separated discipline of economic history, Sweezy has been a prime mover in the debate on the transition from feudalism to capitalism (see Hilton 1978) which has subsequently given way to the 'Brenner Debate' (see Aston and Philpin 1985).

The topics, listed in the previous paragraph and far from exhaustive, are a testimony to the substantial and central contribution to political economy which has been made by Sweezy and those associated with him. The influence exerted upon Marxism and upon all but the most introverted and abstruse of non-Marxist economics has been most powerful. In this context, given that Sweezy's intellectual life now stretches over half a century, it is surprising that little attention has been paid to the origins and evolution of his thought. Of course, it is a commonplace that *The Theory of Capitalist Development* is truer to Marx's *Capital* than *Monopoly Capital*, although the former already contains elements of the theory of monopoly and underconsumption that were to reach their fruition, if not maturity, in the joint work with Paul Baran in which the concept of potential surplus was incorporated.[2]

The purpose of this chapter could hardly be to provide a 'political economy of Sweezy'. For this, the present author cannot claim adequate acquaintance with Sweezy's intellectual and political biography. Indeed, reliance is placed exclusively, if not exhaustively, upon the articles of Sweezy that have appeared in the 'academic' journals together with some of his books. As a study of the evolution of Sweezy's political economy, use has been made then of only a few of the pieces of the jigsaw that make up his intellectual biography, although there is also available some overall picture of the 'final product', as represented in his mature works.

These comments must be borne in mind as the attempt is made to point to and place a neglected work of Sweezy's. This is his study of the British coal industry from 1550 to 1850. This is a significant work but it would be easy to exaggerate its importance. *Monopoly and Competition in the English Coal Trade: 1550–1850* was published in 1938, Sweezy having extended his Harvard doctoral dissertation of 1937, which dealt with the late eighteenth and early nineteenth centuries, to cover the period from 1550 to 1771. What is surprising is that, despite its suggestive title, this book – whether reference to it, its content and even its subject – appears to have been passed over in the subsequent development of political economy associated with Sweezy. Not only does he never refer to it himself but nor do his critics. The work might never have existed but for those few economic historians concerned with the British coal industry for whom it is a standard reference. But it was not written by any old economic historian but by Paul Sweezy and surely this is of significance.

Nor is this explained by some settling of its author's accounts with the past. Groomed and trained in the academic world all, especially Marxists, are forced to make concessions to the requirements of the profession, of which the gaining of a doctorate is a major stepping stone on the way, probably a greater step then than today. But this possibility cannot explain the 'abandonment' of Sweezy's first major work, since it does itself evidence a knowledge of *and* commitment to Marxism. In this it is not, for example, like his note on elasticity of substitution in the first volume of the *Review of Economic Studies* in 1933. For the coal book's first page declares:

My interest in the problems of economic development was first aroused by a study of Marx's brilliant investigations into 'the law of motion of the capitalist system'.

(Sweezy 1938a: ix)

And, having completed an analysis of monopoly and competition in the British coal industry, Sweezy closes as follows:

> Intensive study of a particular sector of the economy first led to the isolation of what appeared to be the main active forces in determining the organization of producers. It was then a natural step to attempt to observe the action of those forces in the economic process as a whole. The conclusions so reached are obviously tentative; they are set forth here, however, because it seems likely that the insight gained by pursuing this method is sufficiently enlightening to warrant a great deal of further study along the same general line.
>
> (p.150)

Such is the task that Sweezy sets himself (and others). A mere four years later in 1942, *The Theory of Capitalist Development* was published. It seems inescapable that it was heavily influenced by the earlier research on the coal industry.

Here, the question is posed of how the book on the coal industry corresponds to and contributes to his later, more general political economy. Clearly, any assessment of the significance of the coal industry research would have to take account of the other influences on Sweezy's thought. As previously mentioned, these cannot all be elaborated, but a few suggestive markers can be placed. To begin with, apart from the classics of Marxism, it is important to recognize that Sweezy is breaking from his training in the orthodox tradition of the early 1930s, in which the formal mathematical models of general equilibrium are emerging into prominence. But even as late as 1949, Sweezy remains attracted by the idea of general equilibrium as a characteristic of Marxist as well as of orthodox economics, referring to:

> what Marx called the 'law of value', which I have elsewhere characterised as a theory or general equilibrium developed in the first instance with reference to simple commodity production and later on to advanced capitalism.
>
> (p. 247)

A second influence on Sweezy was the work of, certainly today, lesser studied writers within the Marxist tradition, in which the issues of underconsumption and economic reproduction loomed large. As Maurice Dobb pointed out in his review of *The Theory of Capitalist Development* in *Science and Society* (1943):

> Of wider interest is the analysis of the various elements in Marx's theory of crises and the discussion of previous writers' contributions to this problem: for example, Tugan-Baranovsky, Conrad Schmidt, Kautsky, Rosa Luxemburg, and Henryk Grossman.

To this influence must be added that of what might be loosely termed the institutionalists, particularly those concerned with the role of the managerial revolution and the modern corporation.[3]

A last influence, and one that combines the others, concerns methodology and, in particular, the relationship between theory and evidence. Sweezy (1934: 800) argues that:

> I take it that the object of economic theory is to frame a set of logically co-ordinated questions, the answers to which are to be sought by consulting the facts.

In retrospect, it is clear that the questions that became central to Sweezy concern the causes and characteristics of monopoly capitalism. The idea emerged that monopolization had dulled the forces of competition and that monopoly pricing is an alternative source of profitability to sustained cost reducing investment programmes. The result, in the context of downward pressure on wages to sustain profitability, is insufficient aggregate demand to allow all potentially available profits (potential surplus) to be realized.

This, in a nutshell, is the theory of monopoly capital as it is finally presented, and much analysis is devoted in Sweezy's later work, most notably with Baran in the classic *Monopoly Capital* (Baran and Sweezy 1968), to studying the effects on the economic system of this simple underlying dynamic. But to what extent is this logic already to be found in Sweezy's earlier work on the coal industry? The second section examines the relationship between the coal study and his later political economy by focusing on the understanding of monopoly and competition. But the first section begins by giving an outline of the British coal industry from 1550 to 1850. An appendix also considers the significance of Sweezy's coal study for the debate over the transition from feudalism to capitalism.

Combination in the coal industry

Until the advent of railways, the transportation of coal was a major factor governing sources of supply to its various markets. As Flinn (1984:

146) observes, 'it was customarily assumed, in the eighteenth century, that land carriage of coal doubled its pithead price in ten miles', but that it could by water be carried 'twenty times as far as by land for the same unit cost.' In the case of the largest domestic market, London, supply was dominated over the period concerned by coast-wise shipments from the North East, so much so that it gave rise to the expression of 'carrying coals to Newcastle' to suggest a futile activity.[4]

This is hardly surprising for the coal trade was a big and expanding business. Between 1586 and 1700 the total quantity of coal imported into London increased from 23,867 to 428,100 tons (Nef 1932, I: 124). By 1830, over two million tons were being shipped to London (Flinn 1984: 217). With London never taking more than 65 per cent of its production, the importance to the North East of the London markets can be exaggerated if they are considered to the exclusion of other markets. For London, however, the North East was the essential source of supply. In 1830, it provided 96 per cent of the capital's coal, this involving 10,000 shipments a year (Flinn 1984: 172). With such a limited source of supply, the coal trade was open to monopolizing activity as an attempt was made to increase price and profit by restricting output.

In principle, these objectives could be achieved by restrictions imposed at any point in the production, transport, and sale of coal in its journey from the seams in the North East to the hearths of the London home or industry. Access to the coal seams was dependent upon the ownership and control of landed property and the same applied to the 'wayleaves' or right of transport to, and use of, the port facilities. From that point to final consumption, there existed a variety of intermediaries ranging from those controlling the colliers that undertook the coast-wise shipments to the wide variety of those engaged in trading, loading, and transporting.

It is the standard history of the industry over the period concerned that an effective monopoly acted from time to time. But, at the outset, it is important to recognize that it could be imposed at various points along the route from production to consumption. Thus, monopoly gains for the economic activities associated with the coal trade could accrue from control over one of its essential component parts. By the same token, the exercise of such control carries with it command over the other activities. To be specific, for example, if there were a monopoly over the shipment of coal, that would carry with it control over who would supply the coal, for entry into the industry as producer may be possible but is futile in the absence of all the conditions necessary to get your coal to market.[5] Consequently, supply itself may be restricted by indirect means:

Landowners had found that the ownership of coalbearing land was not in itself sufficient: access to river or sea was as essential.
(Flinn 1984: 160)

The result was that fifty or more collieries were at one time prevented from working, thereby pushing up the prices in London.

This is important since the origins of the monopoly over the coal trade are not apparently to be found in direct control over supply, but in the Company of Hostmen, formed by Royal Charter in 1600, who enjoyed a monopoly over trading in the town of Newcastle. By 1726, they had given way to the Grand Allies, a combination led by the three main producers. Finally, the Limitation of the Vend, established in 1771, set about regulating competition by ultimately reaching the stage of allocating production quotas to its members with penalties for violation.

Sweezy's contribution to the study of the coal industry focused mainly on the later years of the cartel arrangements. His is an original analysis of the changing form of the monopoly and of its final disintegration in the middle of the nineteenth century. This resulted from the coming of the railway system which opened up alternative sources of supply from a variety of coalfields that could now depend upon overland transport. Sweezy traces the success and failures of the attempts to limit output and maintain prices and he examines the changing institutional structure that evolves to accommodate (and exploit) changing circumstances – whether it be, for example, the changing location of production (as seams are exhausted), methods of transport, or the challenge of combining the interests of producers on the Tyne (Newcastle) and Wear (Sunderland).

In general, subsequent studies have tended to play down the output-restricting, price-increasing role of the Vend. In an econometric model, Hausmann (1980) finds a dummy for its effect on price, when in operation, to be insignificant and even of the wrong sign. In other models (Hausmann 1984a and 1984b), he has found the price effect to be significant but small when compared to the effect of the taxation of coal.[6] But the empirical results of this history are of less concern than the light that its presentation sheds on Sweezy's theoretical outlook.

Monopoly and competition

Sweezy sets out to confirm the relationship between the formation of a cartel and its effect on output and prices. He concludes for the early nineteenth century that:

The *short-run* effect of the Limitation was to keep prices above what they would have been had it not been in existence.
(1938a: 157, emphasis added)

But he continues that:

> Unfortunately, material does not exist . . . on which one might attempt to assess the long-run effect.

This is partly because of the lack of data but also because, after 1850, economic conditions (the arrival of the railways) had changed so much that it would be impossible to isolate the effect of the Limitation.

For Sweezy, in the short-run, the need for cartel arrangements is limited during periods of expansion for:

> whenever an industry is expanding rapidly and existing plant is being used continuously at capacity level there is nothing to fear from cut-throat competition.
>
> (p. 31)

During periods of stagnation, however, there is an incentive to sustain profit by cartel arrangements, for then:

> the great 'natural' profit-preserving force – expansion of the market – bogged down for the time being. The coal owners consequently turned once again to combination.
>
> (p. 143)

Equally, there is in such circumstances the incentive for individual producers to violate the cartel by dropping price and increasing output (pp. 114ff).

It is impossible to fault this outlook, for it simply claims that cartels are only as strong as the commitment to them; that they have an effect on prices and output in the short-run; and that the pressures both to combine and to cheat against a combination are greater the smaller is the size of the market. Not surprisingly, this view is to be found in *The Theory of Capitalist Development*:

> The fact that it is useless to search for a theory of monopoly price which can stand on an equal footing with the theories of value and production price should not, however, be a cause for despair. For it is possible to say with a great deal of generality and assurance that, as compared to the situation which would exist under competition, equilibrium output is smaller and equilibrium price is higher when elements of monopoly are introduced.
>
> (1942a: 271)

Before, however, the transition can be made to the theory of

Monopoly Capital (Baran and Sweezy 1968) two further steps have to be taken – one from the short run to the long run and the other from an individual sector to the economy as a whole. If these steps can be taken, and what is true for the one-sector short run carries over to the economy-wide long run, then clearly the result is that of a stagnating, uncompetitive economy, as found in *Monopoly Capital*. Elsewhere, it has been argued that these steps cannot be validly taken for, unsurprisingly, they require the systematic setting aside of all of the competitive pressures which continue to survive into the era of monopoly capital (Fine and Murfin 1984).

These arguments are not reproduced here. Rather the issue is how Sweezy confronted the significance of his coal study for the economy as a whole. First, in moving from the short to the long run, Sweezy is concerned with competitive conditions:

> The influences which at different times and in varying degrees . . . preserve a margin of profit may be set out as follows:
> 1. Producers may combine and abate their mutual competition for the largest possible share of the market.
> 2. The market itself may expand sufficiently to mitigate or nullify the effect of such competition on selling prices.
> 3. Producers may combine and abate their mutual competition for the available means of production.
> 4. The available means of production may expand sufficiently to mitigate or nullify the effect of such competition on costs.
> 5. Producers already in the field may combine to restrict the entry of newcomers.
>
> (1938a: 136)

Sweezy goes on to argue that capitalist combination involves the use of 1, 3, or 5. But more important is the setting aside of what is for Marxism the most significant aspect of competition in the monopoly stage – the cheapening of commodities through the (machine) production of relative surplus value. Sweezy is certainly aware of this, for he quotes Marx approvingly for recognizing the tendency to monopoly as compared to Ricardo, Mill, Jevons, and Menger:

> 'The battle of competition,' he said, 'is fought by cheapening of commodities. The cheapness of commodities depends, *ceteris paribus*, on the productiveness of labour, and this again on the scale of production. Therefore, the larger capitals beat the smaller'.
>
> (1938a: 148)

Ironically, Sweezy is only interested in the result – larger capitals

– and not the reasoning by which they have been obtained (with one important exception to be taken up at the end of this section). For, 'the growth of large-scale production rendered combination both easy and attractive'. But the cost-reducing effect of large-scale investment has found a more significant role by the time of *Monopoly Capital*. It increases profitability:

> under monopoly capitalism declining costs imply continuously widening profit margins. And continuously widening profit margins in turn imply aggregate profits which rise not only absolutely but as a share of national product. If we provisionally equate aggregate profits with society's economic surplus, we can formulate as a law of monopoly capitalism that the surplus tends to rise both absolutely and relatively as the system develops.
>
> (Baran and Sweezy 1968: 80)

As will be familiar to those acquainted with the monopoly capital school, anything that increases the surplus tends only to give rise to stagnation:

> Twist and turn as one will, there is no way to avoid the conclusion that monopoly capitalism is a self contradictory system. It tends to generate ever more surplus, yet it fails to provide the consumption and investment outlet required for the smooth working of the system. Since surplus which cannot be absorbed will not be produced, it follows that the *normal* state of the monopoly capitalist economy is stagnation.
>
> (p. 113)

Thus, in taking account of the cost-reducing effects of large-scale production, which are neglected in the coal study, Sweezy finds the source of a surplus but not the means to realize or absorb it through exchange.

In moving in his coal study from an individual sector to the economy as a whole, Sweezy situates his approach relative to what is now termed general equilibrium theory. He argues that this should be set aside because it fails to recognize that the forces of competition will be blunted and profit maintained by restricting the operation of market forces:

> The orthodox theory of competitive equilibrium has the great merit of directing attention to those forces which, if unchecked, undermine the profit margin which is the proximate end of business activity. The weakness of orthodox theory lies in its failure to perceive that, so long as capitalism persists, the influences which work to check these forces must on the whole be successful. Failing to perceive this truth, orthodox theory has never analyzed the immensely significant

processes whereby the profit margin is preserved as the mainspring of the capitialist mechanism.

(1938a: 135)

Sweezy then mentions the five factors listed above but what is remarkable about the above quotation is its suggestion that competitive equilibrium would undermine the profit margin whereas, as is well-known, it merely equalizes the rate of profit to its equilibrium level. Perhaps Sweezy is influenced by classical political economy for which the stationary state is the ultimate zero-profit destination of capitalism – recall that he is writing in 1938, long before sophisticated growth and general equilibrium theory, and that much closer to the classical tradition. Indeed, Adam Smith seems to be his mentor for there is occasional reference to the limits imposed by the size of the market, particularly in the coming of large-scale production and its association with monopoly, although there is no reference to Smith's motive force – the growing division of labour – in line with the general neglect of the sources (and effects) of productivity increase:

> The growth of large-scale production rendered combination both easy and attractive; the tendency which it engendered for productive capacity to outrun the market made it absolutely essential.
>
> (p. 148)

Thus, the theory of monopoly capital, as far as combination and cartels are concerned, is already well developed in Sweezy's coal study. Its extension from individual sectors to the economy as a whole and the implications of this for stagnation are present but less well developed. How does this fit into the evolution of his thought more generally?

In reviewing Pigou's *Theory of Unemployment* in 1934, there is no hint of an attachment to the theory of monopoly capital and the most orthodox explanation of unemployment is fully embraced:

> The implication throughout *The Theory of Unemployment* is that, apart from frictional obstructions, unemployment would be non-existent if it were not for the fact that wage-earners habitually stipulate for a rate of wages higher than the 'equilibrium' level. In a theoretical study this is undoubtedly the correct way to approach the problem.
>
> (Sweezy 1934: 807)

But by 1938, the notion that too high wages cause unemployment has been abandoned – 'any one interested in increasing the volume of employment and investment would do well to regard the problem of wage policy as distinctly of secondary importance' (Sweezy 1938b: 157). This conclusion is reached by appeal to the monopolization of the economy and its effects on aggregate demand and through the use of an informal exposition of the 'kinked' demand curves (that were to become a standard reference through the article published later in the *Journal of Political Economy*, Sweezy (1939)):

> Let us suppose, for example, that all employers are monopolists or, as it would probably be more accurate to say, oligopolists. In such circumstances the demand curves which are relevant to behavior are those which exist in the heads of the entrepreneurs, and they are made up after a full calculation, frequently more or less unconscious, of the probable reactions of more or less distant rivals. I am prepared to argue that a very likely shape for such demand curves will show a 'corner' at the prevailing price. The entrepreneur in other words is likely to regard price cutting as self-defeating since it provokes similar action on the part of rivals (the curve is inelastic for downward movements of price).
> (Sweezy 1938b, 156)

In this contribution, Sweezy is informally anticipating the contemporary concern with oligopolistic behaviour and equilibrium under conditions of entrepreneurial conjecture about rival reaction. The same concern arises only in passing in terms of 'group' equilibrium in a 1937 note in the *Quarterly Journal of Economics* which attempts to define monopoly and this contains little analytical content. During this period, Sweezy would have been working on his coal study and, not surprisingly, the problem of group equilibrium is reproduced in the *JPE* article, but this leads to a further conclusion that many equilibrium solutions are viable in the context of the oligopolistic economy:

> it becomes very doubtful whether the traditional search for 'the' equilibrium solution to a problem in oligopoly has very much meaning. Generally speaking, there may be any number of price–output combinations which constitute equilibriums in the sense that *ceteris paribus*, there is no tendency for the oligopolist to move away from them. But which of these combinations will actually be established in practice depends upon the previous history of the case. Looking at the problem in this way the theorist should attempt to develop an analysis which will enable him to understand the processes of change which characterize the real world rather than waste his time in chasing the will-o'-the-wisp of equilibrium.
> (pp. 572–3)

Three interesting points emerge from this. First, the rejection of a determinate equilibrium is liable to lead Sweezy to set aside Marx's value theory in so far as he perceives it to be a general equilibrium for competitive capitalism, as noted earlier. Second, that equilibrium is indeterminate means that full employment for the economy as a whole is a special case and, in general, oligopolists will be holding up prices and holding off from expanding output. Third, there is, of course, a parallel here with Keynes' idea that Walrasian full employment equilibrium is a special and extreme case in which there is adequate effective demand.

Sweezy's subsequent attempts to characterize monopoly capital can be viewed as arguments to support the proposition that the special, extreme case, arising from the last two points, will not occur. For example, in discussing the 'managerial revolution', Sweezy (1942b: 21) argues that 'monopoly replaces competition, and the limits of accumulation narrow relatively to the needs of capital.' More clearly, in discussing Schumpeter's theory of innovation, Sweezy (1943) is clearly aware of the coercive pressures to accumulate:

> First, accumulation in the absence of change tends to wipe out the surplus and hence to threaten the social position of the capitalist; needless to say, no capitalist tamely submits to his own extermination. Second, the individual capitalist who introduces new methods makes a larger surplus and hence can get ahead more rapidly than his fellows. Finally, the capitalist who refuses to enter the race for new and improved methods stands in danger of being eliminated by his more alert competitors.

At this point, Sweezy refers readers to his *The Theory of Capitalist Development*. But, on the next page, the relevance of coercion to accumulate for an era of monopoly capital is brought into question for 'in the case of a modern large-scale corporation or combine in which the process of innovation becomes highly institutionalised in the hands of staffs of research scientists, cost accountants, etc., and in which it might be extremely difficult if not impossible to find a Schumpeterian entrepreneur'. From this point to *Monopoly Capital*, there is greater sophistication and substance of argument, but the essential features of a stagnating, oligopolistic economic system are fully present.

Concluding remarks

This chapter has tried to show that Sweezy's coal study contains many, if not all, of the propositions that were going to make up the basic postulates of *Monopoly Capital*. Certainly there is the neo-Smithian concept of the importance associated with the limits of the market and also there is the post-Keynesian building up of monopoly capitalism from the price-increasing, output-restricting behaviour within individual sectors. Further, it appears as if the economy is understood in terms of its points of departure from a competitive equilibrium whose 'natural' resting place is profitless. In this light, it is hardly surprising that monopoly should be seen as a (vain) attempt to stave off declining profitability and that other economic and social phenomena (such as arms expenditure, advertising, etc.) should be located in terms of futile countertendencies to the normal state of stagnation.

For Sweezy, combination in the coal industry at the beginning of the nineteenth century was anomalous as capitalism was entering its period of classic *laissez faire* competition with the stage of monopoly capitalism beyond the horizon a century ahead. He explains this monopoly in terms of a coincidence of locational, economic and social factors (1938a: 147). Despite this, however, it can be seen that his method of understanding this monopoly, prior to the monopoly stage, translates easily into the analysis for the later period. As a result, the criticism by Weeks (1981) of Baran's and Sweezy's projecting monopolies of the past on to the modern world has some support. It is also remarkable that the two objects of study, the old coal industry and the new era of monopoly capitalism, should each have experienced extensive periods of expansion for the times that they were considered to be under monopoly, output-restricting conditions.[7]

This suggests that the traditional account of the coal industry for the period concerned might be opened for reconsideration, just as has been the analysis of *Monopoly Capital*. After all, the notion of combination as the central feature of the coal trade takes no account of the organization and changing relations of production, even over a period of three hundred years. But this is not surprising when monopoly as market control alone is taken as the central aspect of the industry.

Appendix 1.1: The transition to capitalism

The substance of this chapter tends to suggest that at an early stage Sweezy fell into a neo-Smithian understanding of the transition to the capitalist economy as has been powerfully argued by Brenner (1977) for his later work. More specifically, the transition debate concerns the issue of whether the dissolution of feudalism was more brought about by its own internal contradictions (Dobb) than by the pressures of external commerce (Sweezy). Following on from this are a number of subsidiary issues of which the most important here are the origins of capitalists and the role of towns, although we shall consider only the former of these. (See Hilton 1978: 13.)

But a prior question is whether the coal trade between 1550 and 1850 is relevant to the transition debate. There are good reasons to see why it should have been considered irrelevant by Sweezy. For him the transition should already have taken place because of his identification of capitalism with commerce, 'since trade can in no sense be regarded as a form of feudal economy' (1978: 40). A worthwhile monopoly could then hardly have been countenanced as having relevance for the debate – even more so for the coast-wise shipping involved:

> we can define western European feudalism as an economic system in which serfdom is the predominant relation of production, and in which production is organised in and around the manorial estate of the lord. It is important to notice that this definition does not imply 'natural economy' or the absence of money transactions or money calculation. What it does imply is that markets are for the most part local and that long-distance trade, while not necessarily absent, plays no determining role in the purpose or methods of production. The crucial feature of feudalism in this sense is that it is a system of *production for use*.
>
> (p. 35)

Of course, this restrictive definition of feudalism is open to dispute and Hilton (1978: 30) provides a corrective. But even though Sweezy might have felt the coal industry was irrelevant in establishing his own case, he might at least have turned it against his opponents by suggesting that their approach would have unreasonably led to the conclusion that the coal industry would have to be considered feudal on their definition. Instead, he set aside the opportunity of drawing upon what is probably his only extensive primary source research, 'like his opponent in this controversy, Paul Sweezy was in a similar situation – that is, of a Marxist analyst of contemporary capitalism who ventured into the field of medieval economic history on the basis of secondary work by non-Marxist historians' (Hilton 1978: 12).

There are, however, other reasons why Sweezy may have set aside his own research in order to defeat his opponents. First, for the coal study to have been an effective weapon, he would have to have entered the terrain of his opponents. For them, the definition as well as the moving force of feudalism depended on the relations of production. These and conditions of production more generally are scarcely examined by Sweezy. Just before the end of the period covered by his study, miners were in conflict with the owners over the terms and conditions and the very existence of the annual bond – evidence enough that the freedom of the workforce to sell their labour power had yet to be completely won. But for Sweezy, such conflict is discussed more in terms of distribution and the pressures placed upon the stability of the coalowners' combination.

For the period from 1550 to 1771, where the issue of transition in the relations of production is more relevant, however, Sweezy relies upon the work of Nef and does not himself discuss labour. Nef's own approach appears to preclude the transition as bearing on the relationship between feudalism and capitalism. He finds substantial evidence of slavery amongst the Scottish workforce but otherwise he only categorizes labour and searches by the criteria of the presence of either the wage system or of independent combinations of workmen (whose presence and survival was precarious). In short, while no conclusion is presented here on the significance of the coal industry in this period for the transition debate, it can be seen why its use by Sweezy would have been difficult and unhelpful.[8]

Second, however sharp the weapon of the coal industry, it is unlikely that Sweezy would have been able to turn it against his opponents. On the issue of whether capital evolved by the revolutionary route of producer entering the sphere of exchange (Dobb) or by merchants entering production (Sweezy), it might be thought that the hostmen, as a trade monopoly, were evidence of the move from commerce to industry. Indeed, this is the argument of a study of the Tyneside ruling class, apparently unaware of Sweezy's coal study:

> Whilst the old land-owning aristocracy from the 1800s onwards played no significant part in this process of industrialisation, the role of a pre-existing merchant class was particularly important throughout the period. This is clear from the evidence of their considerable and direct involvement as entrepreneurs in the more highly capitalised sectors and suggests that the view that merchant capital did not play a progressive role in developing industrial capitalism does not hold for all areas and all periods and needs some qualification.
> (Benwell CDP 1979: 16).

This is highly supportive of Sweezy and the qualification is intended to apply to a quotation taken from Marx:

> Wherever merchant's capital still predominates we find backward conditions. This is true within one and the same country, in which for instance, the specifically merchant towns present far more striking analogies with past conditions than industrial towns.
> (Quoted in Benwell CDP 1979: 16)

Such conclusions are quite impossible for Sweezy on the basis of his coal study because, for him, the coal producers were always in command:

> It is important to understand that in 1600 it was the big *mine owners*

The coal question

> who controlled the municipal government of Newcastle and the Company of Hostmen, in spite of the fact that no exclusive privileges with respect to owning and working collieries existed . . . . The history of the next century is the history of the separation of the mine-owning hostmen from the rest of the fraternity, a process which took place along with the gradual wearing out of the Company's special privileges.
>
> (1938a: 22)

In other words, and Sweezy sharply disagrees with Nef in so arguing, the producers controlled the coal trade, and the rise and fall of merchant capital in terms of the fortunes of the Company of Hostmen merely represented different ways of their doing so. This could hardly be fertile ground for Sweezy in his debate with Dobb. It is perhaps unfortunate, in this context, that the transition debate has in its Brenner phase concentrated so closely on agriculture. But the significance for this debate of a re-interpretation of the coal industry is beyond the scope of this chapter.

Chapter two

Cartels and rationalization in the 1930s

In chapter 6, it will be shown that the statistical evidence points to the presence of economies both of scale and of mechanization in the coal industry in the 1930s. Yet the industry was extremely slow to take advantage of these, remaining a highly labour intensive and fragmented industry, particularly when compared with its competitive rivals for export markets, and backward technically and in mine layout, as extensively documented in the Reid Report (1945). This leaves open the question of why this should have been so, unless the blame is simply and fully laid at the door of the mine owners. This has not been an unpopular explanation. For it is to Lord Birkenhead that is owed the classic statement in criticism of the mine owners' entrepreneurship:

> it would be possible to say without exaggeration that the miners' leaders were the stupidest men in England if we had not had frequent occasion to meet the owners.

But there are other explanations for poor performance, including low levels of (export) demand, poor industrial relations, unsuitable geology, limited availability of finance, and insufficient attention to coal preparation and marketing. None of these, however, automatically translates cause into effect in the sense of unambiguously being associated with low levels of mechanization and rationalization. It is simply assumed that one form of backwardness or difficulty gives rise to another. But poor industrial relations, for example, might easily be thought to give rise to the closure of troublesome pits and a greater ease of concentration of production.

It is against these general observations that the role of cartels is assessed in the performance of the industry in the interwar period.[1] The notion that a state-organized cartel in the 1930s featherbedded inefficient producers, thereby impeding amalgamations, is critically assessed. The first section theoretically rejects the idea that low pace of amalgamations is the necessary product of cartels. The second section overviews the evidence concerning amalgamations and finds that its rhythm seems to have been more in line with the level of demand than with the presence or not of a cartel. The third section questions in any case the extent to which the cartel was in fact successful leading to the conclusion that the causes of the industry's problems have to be sought elsewhere. Although it is not suggested that the cartel is insignificant, it is perhaps better to see it as a consequence of the weakness of the forces for reorganization rather than the cause.

Featherbedding cartel?

One standard explanation for the failure of the industry has been put forward by Kirby (1973a, 1973b, and 1977). It concerns the supposedly inconsistent effects of the Coal Mines Act of 1930 and so applies only to retardation of the industry in the 1930s. Under Part II of the Act, the Coal Mines Reorganization Commission (CMRC) was set up under the Chairmanship of Sir Ernest Gowers (Thomas 1937). It was charged with the task of reorganizing the industry into a smaller number of larger units. It was given powers of compulsion, subject to the judgement of the Court of the Railway and Canal Commission, on whether the amalgamations would lower the cost of coal, be in the public interest, and accord with the financial interests of those mine owners directly concerned.

The CMRC, despite violent opposition from the mine owners, was determined to push ahead with schemes for amalgamation. It considered voluntary action preferable but was prepared to use its powers of compulsion. It was eventually obliged to do this in the face of the intransigence of the mine owners in co-operating with any of the schemes of amalgamation that were proposed. But in 1935 a test case was heard before the Court whose decision made it clear that the powers of compulsion held by the CMRC under the Act were, in practice, non-existent. The activities of the CMRC were effectively suspended, for on 28 July 1935 the Secretary for Mines asked it 'to refrain from initiating fresh inquiries in regard to possible amalgamations pending consideration by the Government of the whole position and powers of the Commission.' All the CMRC could do was to work behind the scenes to try and obtain greater legislative powers. These were not granted in any meaningful way and the reorganization of the industry had to wait upon nationalization in 1947.

So much for Part II of the Act. Part I was intended to be used to set up a state-organized cartel for the industry. Statutory marketing schemes, which were to come into operation on 1 January 1931, were designed to mitigate the structural adjustment costs of a declining industry by discouraging 'cut-throat' competition. Consequently, the schemes were intended to:

improve the finances of the industry as a whole, and thus give the work-people increased wages and the owners a reasonable return on the capital invested.

(Board of Trade 1936)

In other words, the idea was to hold up prices by holding down output. Indeed, the intention, as will be seen, was to allocate a limited output to each colliery company.

Kirby argues that these two Parts of the Act were inconsistent with each other. Whilst Part I was designed to bring about reorganization, Part II would frustrate this by maintaining profitability in the otherwise uneconomic pits. Specifically:

> the statutory cartel system, in preserving the existing structure of the industry, lessened the incentive to amalgamate.
>
> (1973a: 282)

This is by no means a novel argument. It was even used by a commentator on the industry at the time:

> quota works in the majority of cases to swell the value of inefficient undertakings, and has the effect of retarding amalgamations.
>
> (Neumann 1934: 343)

This view is strongly disputed here, at least in the simple form in which it is posited. At the theoretical level, there is no reason to presume that a cartel will hinder amalgamation. For a cartel is concerned to achieve higher prices for any given level of output. Whatever levels of profits this allows, they can be made higher the more efficiently or less costly are the methods by which the output is produced. Indeed, cartels are often seen as the basis on which rationalization of industrial structure can proceed and the German coal industry of this very period is held up as an example. The point about a cartel is that it may diminish the coercion to bring about reorganization, through monopoly pricing sustaining profitability, but it certainly does not diminish the incentive to cost reduction which is, in principle, a separate source of profitability.

In short, it is impossible to explain the development of an industry through the existence of a cartel, although this may have an important if not a determining influence. For the cartel is consistent, through lack of coercion and presence of incentive, respectively, with either ossification or reorganization.[2] Some other factors must be brought forward to explain which of these two outcomes (or somewhere in between) actually is brought about. Otherwise the theory is, in retrospect, foolproof. If rationalization occurs, as in Germany, it would be argued that incentive to cost reduction prevailed. If not, it would be said that coercion failed, as in the British case. But this does not explain why coercion was the prime mover in one instance and incentive in the other. Nor can these general propositions concerning the effect of cartels, monopoly, or whatever provide an answer. Rather, this depends upon identifying the contingent forces promoting or impeding reorganization. These arguments are taken up in greater detail in Fine and Murfin (1984).

Amalgamations: the empirical evidence

In short, Kirby's theoretical approach is flawed. However, he also brings forward empirical arguments to support his position but, as will be shown, these are equally flawed. In demonstrating this, the earlier theoretical remarks will be confirmed. Kirby tries to show that prior to 1930 the pace of amalgamation was greater than afterwards, illustrating the obstructive effect of Part II of the Act. From 1926 until 1930, under the Mining Industry Act of 1926, voluntary amalgamations were sought. Also, the Act conveniently made the provision for data to be collected by the Board of Trade on the amalgamations effected throughout the remainder of the interwar period. The data are summarized in table 2.1, which includes the rather scanty evidence used by Kirby. He notes that there were 26 amalgamation schemes between 1926 and 1930 compared with 32 in the longer period from 1930 to 1936. The number of workers involved in aggregate was 212,260 and 164,200, respectively. Even if this were all there were to the matter, this hardly represents a dramatic decline and is more evidence of how limited amalgamations were in both sub-periods.

But there are major deficiencies in Kirby's analysis even before account is taken of the weakness and limited amount of evidence that he considers. First, it is important to recognize that an amalgamation is defined as the complete unification of ownership of two or more undertakings. As a result, the number of pits and workers associated with each merger is somewhat misleading. Specifically, these numbers are found by simply adding together the number of pits and workers of each undertaking involved in the merger and this may lead to a certain degree of double counting over time.

For example, if a dynamic enterprise absorbed new collieries each year, the number of pits and workers involved in this merger activity would steadily increase. On the other hand, the number of pits available for new mergers would tend to decline as they are absorbed, unless there is frequent entry of new pits into the industry.

By way of illustration, the Powell Duffryn Colliery Company was formed in 1928 in South Wales, involving 26 pits and 25,100 workers. This left fewer pits and workers available for subsequent amalgamations.

Cartels and rationalization in the 1930s

Table 2.1 Amalgamations in the coal industry 1927–38

Year	Schemes	Pits	Employment	Employment per scheme[1]	Pits per total[2]	Total output[3]
1927–						251
1928	17	172	125,960	7,409	3.19	237
1929	6	61	39,800	6,633	2.52	258
1930[4]	3	88	46,520	15,507	3.78	244
1931	3	28	10,500	3,500	1.25	219
1932	4	27	12,390	3,096	1.25	209
1933	6	22	19,580	3,263	1.03	207
1934	2	28	12,600	6,300	1.32	221
1935	8	173	83,100	10,388	8.34	222
1936	7	43	22,750	3,250	2.07	228
1937[5]	18	95	83,253	4,625	4.48	240
1938	10	46	24,726	2,473	2.16	227
		Subtotals				
1927–30	26	321	212,280	8,165		
1931–6	30	321	160,942	5,365		
1927–30	26	321	212,280	8,165		
1931–4	15	105	55,070	3,671		
1935–8	43	357	213,829	4,973		
		Subtotals (Kirby)[6]				
1926–30	26		212,260			
1931–6	32		164,200			

Source: Reports by the Board of Trade Under Section 12 on the Working of Part I of the Mining Industry Act, 1926 (Reports 1–11).
Notes:
1. Column 4 divided by column 2.
2. Pits involved in amalgamations expressed as a percentage of the total number of working pits.
3. Millions of tons.
4. Dominated by a single entry in South Wales with 60 pits and 32,000 workers.
5. The number of pits and employment are understated.
6. Kirby (1973a). The minor discrepancies between Kirby's and these figures remain unexplained.

In the limit, merger activity would have to cease once a monopoly had been established! Even so, in 1935, Welsh Associated Collieries were also absorbed, yielding a total involvement of 93 pits and 37,600 workers, many of which were already measured in the first amalgamation mentioned. In short, it is by no means clear how well these figures measure the rhythm of amalgamation when they are aggregated in the way that they have been by Kirby.[3]

Second, leaving this aside, from an examination of the number of mergers occurring year to year it appears that initially this is relatively high; begins to diminish; and eventually recovers, surpassing the initial levels. Three different phases can vaguely be discerned. The years 1927–30 (26 combinations) represent the period independent of the

The coal question

effects of the statutory cartel regulations. The second phase, covering the years 1931–4 (15 combinations), corresponds to both the introduction of statutory cartels and the depths of the 1930s depression. Total production of coal falls to its minimum levels during this sub-period. This curtailment of production arose from demand deficiencies rather

Table 2.2 Amalgamations in the coal industry 1927–38
Regional evidence

	Number of schemes										
Region	1927–8	1929	1930	1931	1932	1933	1934	1935	1936	1937	1938
1	6	0	1	0	1	0	0	1	0	3	0
2	8	0	1	0	0	0	0	1	0	3	2
3	2	1	0	2	0	0	0	0	0	6	2
4	1	2	0	0	1	1	1	0	1	0	1
5	0	3	1	0	1	1	0	1	0	0	0
6	0	0	0	0	0	1	0	0	1	0	0
7	0	0	0	0	1	0	0	0	1	2	1
8	0	0	0	0	0	1	0	0	2	1	0
9	0	0	0	0	0	0	0	0	0	0	0
10	0	0	0	0	0	0	0	1	0	0	0
11	0	0	0	0	0	0	0	0	1	0	0
12	0	0	0	0	0	0	0	1	0	1	0
13	0	0	0	0	0	1	0	0	0	0	0
14	0	0	0	0	0	0	0	0	0	0	0
15	0	0	0	1	0	0	1	1	1	0	0
16	0	0	0	0	0	1	0	1	0	2	3
17	0	0	0	0	0	0	0	0	0	0	0
18	0	0	0	0	0	0	0	0	0	0	0
19	0	0	0	0	0	0	0	1	0	0	1

Source: Reports by the Board of Trade Under Section 12 on the Working of Part 1 of the Mining Industry Act, 1926.

Regional code
1. South Yorkshire
2. South Wales
3. West Yorkshire
4. Northumberland
5. Lancashire and Cheshire
6. North Wales
7. Durham
8. North Derbyshire and Nottinghamshire
9. South Derbyshire
10. North Staffordshire
11. Cannock Chase
12. Warwickshire and South Staffordshire
13. Leicestershire
14. Cumberland
15. Ayrshire
16. Lanarkshire
17. Lothians
18. Fife
19. Somerset

than from stringent output regulation. The final phase, covering years 1935–8, is associated with a brief recovery in demand, during the years 1936–8. Significantly, the number of amalgamations jumps to 43 for the years 1935–8. By fortuitous accident or deliberate design, Kirby stops short his empirical analysis in 1936, just when the merger phase appeared to be reaching a peak and, thereby, completely discrediting his arguments.

Third, it is revealing to examine the rhythm of merger activity across the separate districts. The argument used for the UK as a whole should apply equally to each district. Data are presented in table 2.2. What this shows is that only five districts were involved in merger activity in the first phase, 1927–30; South Yorkshire, West Yorkshire, South Wales, Northumberland, and Lancashire and Cheshire. Subsequently, merger activity seems to have been more widely dispersed across the districts during the period, for Kirby, when it should have been slackening off. His limited aggregate evidence may be more of a statistical artefact reflecting diverse movements in different districts for reasons unrelated to the rhythm of cartel formation.

To summarize; the empirical evidence on amalgamations, considered both fully and in detail, does not appear to support the idea of an association between the Coal Mines Act of 1930 and an impediment to industrial reorganization. But this is not the end of the story.

The cartel in operation

The previous section has shown that the effects of the cartel predicted by Kirby appear to be absent. But this is all on the presumption that the cartel did indeed operate effectively. Was this the case?

It can be argued that Kirby has treated the role of cartels too casually both before and after the 1930 Act. Before 1930, there were a number of voluntarily formed cartels. There were three principal regional cartels, located in South Wales, Scotland, and the Midlands; each was formed around 1927–8 and covered the vast majority of each district which together were responsible for approximately 80 per cent of total national output. No inter-regional co-ordination was established.

The South Wales regional cartel was the first to regulate output and prices. Two separate schemes were introduced. The first of April 1928 sought to steady the market by stabilizing prices. Thus, minimum price schedules were established and price cutting was penalized. The second scheme of autumn 1929 attempted to regulate both production, via quotas, and prices. Although the price of Welsh coal was steadied, competition from 'outsiders' helped undermine the scheme. The subsequent statutory cartels were based upon this latter scheme.

The Midland counties formed the Central Collieries Commercial Association (better known as the Five Counties Scheme). This cartel included Yorkshire, Nottinghamshire, Derbyshire, Lancashire and Cheshire, Cannock Chase, Leicestershire, North Staffordshire, and Warwickshire. The object of this selling association was to regulate output and encourage exports. The scheme was implemented in April 1928 and lasted 18 months. Quotas generally were not exceeded (perhaps due to their generosity), price competition continued unabated, and exporters were able to increase their share of the market (largely at the expense of other domestic exporters).

The Scottish Coal Marketing scheme was in operation for a year beginning in May 1928. This scheme, supported by 90 per cent of colliery owners, sought to raise prices and keep inefficient collieries idle. The high cost of compensation for idle capacity contributed to the demise of this scheme. These three voluntary regional schemes enjoyed some limited success prior to the 1930 Act but, in retrospect, they are perhaps more significant in showing how difficult it was to hold a scheme together against the pressures of market competition in stagnating markets.

No doubt this gave an enormous impetus to the support by mine owners for Part I of the Act, as compared to their outright opposition to Part II. But it cannot be presumed that the provisions of Part I were immediately effective, as is implicit in Kirby's analysis of their impeding effect on amalgamation from 1930 onwards. As noted on p. 21, a principal object of the marketing schemes was to improve the finances of the industry. Given that the demand for coal continued to deteriorate (further undermining the revenues of the industry) the incentives to violate cartel regulations cannot be overemphasized. It is essential to examine the available evidence regarding the performance of the statutory cartel to see to what extent it could have shielded the inefficient from competition by maintaining prices.

The performance of these marketing schemes are outlined in *The Reports of the Board of Trade Under Section 7 of the Coal Mines Act 1930, on the Working of the Coal Selling Schemes Under Part I of the Act*. The Board of Trade reported eleven times during the years 1931–8. Briefly the schemes were administered using two tiers. The nation was divided, ultimately, into seventeen regional districts (each with its own regional Executive Board); these represented the lower tiers. The top tier was the Central Council of Colliery Owners. The regulation of output was nationally organized by the Central Council but the responsibility for minimum prices was left to each District's Executive Board. Hence, the control of prices was not co-ordinated between districts; further, there was no compulsion for Executive Boards to set economically meaningful minimum prices. Indeed, the Scottish Executive Board was the object of many complaints from adjacent districts, for failing to set restrictive minimum prices and, subsequently, dumping their output. As for the regulation of output, each mine was assigned a standard tonnage, the

Table 2.3 Cartelization of the coal industry 1931-8
Regulation of output

Output: million tons

Year	Allocation[1]	Actual[2]	Difference	Supplement[3]
			%	
1931 Q1	62.586	58.378	−6.75	
Q2	58.216	54.878	−5.75	
Q3	55.132	52.169	−5.37	
Q4	61.518	58.807	−4.41	
Total	237.450	224.230	−5.57	
1932 Q1	67.100	56.988	−15.07	
Q2	56.569	53.424	−5.56	
Q3	55.002	47.285	−14.03	
Q4[5]	57.515	55.807	−2.97	
Total	236.186	213.500	−9.60	
1933 Q1	57.520	55.810	−2.97	
Q2	53.740	49.350	−8.17	
Q3	49.017	48.136	−1.80	
Q4	57.772	56.925	−1.47	3.4
Total	218.050	210.221	−3.59	
1934 Q1	61.680	61.003	−1.10	4.66
Q2	55.298	53.652	−2.98	4.99
Q3	53.606	52.348	−2.35	3.65
Q4[4]				
Total	170.580	167.003	−2.10	4.45
1935 Q1	61.794	59.425	−3.83	1.34
Q2	55.379	54.173	−2.18	4.52
Q3	54.284	53.132	−2.12	3.57
Q4[4]				
Total	171.457	166.730	−2.76	3.07
1936 Q1	64.172	62.483	−2.63	2.65
Q2	57.210	54.847	−4.13	0.31
Q3	57.168	55.477	−2.96	4.59
Q4	63.056	61.362	−2.69	3.06
Total	241.610	234.170	−3.08	2.66
1937 Q1	63.270	61.409	−2.94	1.26
Q2	64.018	62.359	−2.59	9.46
Q3	62.835	59.063	−6.00	1.37
Q4	65.946	64.272	−2.54	3.48
Total	256.070	247.100	−3.50	3.91

The coal question

Year	Allocation[1]	Actual[2]	Difference	Supplement[3]
			%	
1938 Q1	67.410	64.614	−4.15	0.02
Q2	62.320	55.734	−10.57	0.04
Q3	55.153	53.933	−2.21	3.62
Q4	62.221	60.458	−2.83	9.10
Total	247.100	234.740	−5.00	3.12

Source: Reports of the Board of Trade Under Section 7 of the Coal Mines Act, 1930.
Notes:
1. Coal allocated by the Central Council to Districts (aggregated) under Part 1 of the Coal Mines Act, 1930.
2. Output as returned by the Executive Boards to the Central Council.
3. Supplement is expressed as a percentage of the total allocation, e.g. 100 × (supplement/final allocation).
4. Data for December Quarters 1934 and 1935 are not reported by the Board of Trade.
5. During the period 1932 Q4 the Central Council altered the principle on which the quotas were determined.

proportion of which allowed to be produced being determined by the district quota, set nationally by the Central Council.

Two phases of the statutory cartel can be usefully distinguished, these being the periods before and after the introduction of nationally co-ordinated centralized selling (made operational during 1936). During the years 1931–6 the Central Council adopted two broad methods of determining quotas. The first involved setting the quota, in advance, as some proportion of the allocation for the corresponding quarter of the previous year. As demand was in decline, actual production continually fell so that discrepancies between allocations and actual production were relatively large, as shown in table 2.3.

To reduce this disparity the Central Council in late 1932 began setting much more restrictive quotas, but encouraged districts who required additional allocations to meet identifiable demands to apply for more. Hence, from the table, it follows that percentage differences between allocations and actual production fell. This does not imply that the supply of coal was demand constrained. Specifically, additional applications were, it seems, readily obtainable; in fact, during this period many districts successfully applied for extra quotas but did not use them. To illustrate, during the fourth quarter of 1932 all districts except Durham and Cumberland received additional quotas (South Staffordshire received three increases); during the first quarter of 1933, 14 applications were made, all successfully; in the second quarter of 1933 demand fell making additional output quotas unnecessary for most districts, however, 5 applications were made; 4 were successful.

From late 1934 the Central Council began distinguishing supplies for inland and export markets, with quotas being separately assigned. On the whole, export quotas were overly generous, based on the hopeful notion of capturing export markets. Although the reports typically suggest output regulation proceeded smoothly and worked well, this presumably is in terms of minimizing the discrepancies between allocations and actual production without effectively restricting supply. Table 2.3 shows, for the end of 1933 onwards, the supplementary allocations expressed as a percentage of the total. From the table it emerges, at least on an aggregate level, that quotas exceeded supply and that supplements were quite generous. In short, for 1931–6, it seems that additional quotas were granted if, for example, demands were identified. Clearly, such practices are inconsistent with an object of restricting supply to push up prices.

Concerning the regulation of prices during the period 1931–6, it must be concluded that the schemes were also unsuccessful. All districts except Scotland introduced a set of minimum prices by the beginning of January 1931. Scotland, under pressure from the Central Council, introduced its set of minimum prices in August 1931. Scotland provoked much discontent amongst its surrounding districts, and highlighted the principal weakness of locally determined price structures, by establishing prices which, it was alleged, bore no relation to market conditions (i.e. the price constraint was slack). This lack of co-ordination of prices was heavily criticized throughout the period, especially by districts losing their local markets to those predatory neighbours who enjoyed a greater degree of slack in their minimum price schedules. The Central Council recognized from the outset that the lack of co-ordination of inter-district minimum relative prices would be troublesome and sought during the period to tighten up the regulation of prices. The Report of the Board of Trade for the year ending 1934 acknowledged that the lack of cooperation fostered suspicion and provided added incentives to undermine price regulations.

Yet another factor, not necessarily weakening the cartel, but preventing it from acting as a barrier to amalgamation, was the trading in quotas (see also Lucas 1937). These could be bought and sold between colliery companies so that, in principle, there was no obstacle through demand to an expanding colliery, as long as the cost of a unit of quota lay below the reduced costs of some form of rationalization. The Board of Trade reports considerable use of the system of quota transfer, with exchange at a few pence per ton. In South Wales, for example, in the last quarter of 1931, 700,000 tons of quota were transferred in 108 transactions out of a total allocation of 11.1 million tons and an actual production of 10.1 million tons. Only in Scotland do transfer prices appear to have been high with complaints that they reached one shilling and sixpence per ton.

The second phase of the statutory marketing schemes covers the period from the fourth quarter of 1936 to 1938. It was during this period that the revised provisions for inter-district co-ordination of relative minimum price schedules were established. The revisions were intended to meet three conditions. First, all colliery owners must be covered. Second, inter-district competition must be suppressed. Third, evasions of the regulations must be prevented. For most regions co-ordination was achieved through informal meetings of salesmen and regular meetings of joint Executive Board Committees. The revised district selling schemes took three forms. First, the colliery owners of Shropshire, South Staffordshire and the Forest of Dean sold their output to their respective Executive Boards who then marketed the coal. Second, the Midlands (Amalgamated) District divided itself into groups which then sold its output to the Executive Board. Third, the remaining districts adopted a 'Central Control of Sales' scheme, where colliery owners continued to market their own coal but were constrained by sales permits issued by the Executive Boards Sales Committee. These permits typically defined quantities to be sold, prices charged, commissions, subsidies, discounts, and even terms of credit.

With regard to regulation, the price of coal was successfully stabilized during this period. However, even the Board of Trade did not attribute this strengthening of prices to the operation of the revised selling schemes (by the reduction of inter-district competition) but rather to expanding demand. Specifically, concomitant with the revised selling schemes was an improvement in demand conditions. Thus, the Board of Trade argued that the strong increase in demand facilitated the stabilization of coal prices. This expansion in demand continued until the first quarter of 1938, but it contracted during the subsequent two quarters and began to increase by the fourth quarter. Since prices did not collapse during the downturn in demand the Board of Trade argued that the inter-district co-ordination did suppress competitive pressures, so avoiding ugly inter-regional price wars. Hence, to a limited extent, the new revised selling schemes were judged successful during the second phase, after 1936, under the new tighter regulations but were proved for a six-month period only. The effectiveness of price regulation cannot be assessed reliably over such a short period.

Nor were the earlier years of this second phase particularly successful. During 1936 the Secretary of the Mines relaxed certain features of the selling schemes to allow producers to respond promptly to increases in demand, thus avoiding bureaucratic delays in the processing of permit applications. By the end of 1937 the Board of Trade was driven to state (Cmd. 5773):

> The margin by which allocations exceeded performance and the number of supplementary allocations granted, indicated the extent to which quantitative control was temporarily relaxed. It can indeed be

said that during 1937 such control was virtually non-existent, and that any shortages of supplies were not due to the operation of the schemes.

Similarly, during 1938 the Board of Trade again suggested that the quantitative regulation of output had at times been inoperative in practice.

Concluding remarks

Theoretically, it has been suggested that cartel-like behaviour does not of itself either encourage or discourage amalgamation. To summarize, in addition, the empirical arguments against Kirby's hypothesis; there is no evidence to support the idea that the state-organized cartel of the 1930s impeded amalgamations. This is because amalgamations may have slackened in the early 1930s but they picked up in the later 1930s. This seems to reflect the state of the demand for coal. Moreover, this rhythm of merger activity was not uniform across districts. Further, cartels existed prior to 1930 and were, in any case, at their weakest in the early 1930s when merger activity was lowest, and only strengthened, if at all, in the late 1930s when merger activity was at its height.

Interestingly, Henley (1988) does find a significant increase in mark-up over costs for the 1930s coal industry, estimating it at an increase of 12 per cent over earlier levels for the UK as a whole, taking account of factors such as mechanization and variation in demand. However, he fails to acknowledge fully the weakness of the cartel, as elaborated in detail here, nor does he take it into account statistically even though the drift up in the mark-up over the 1930s reflects a combination of ultimately rising demand and a more powerful cartel (after 1936) as well as his preferred process of entrepreneurial adjustment to the new cartel-based equilibrium. In addition, as he acknowledges, combined within the mark-up will be the results of a weakening of labour's position (reflected in wages costs) following the General Strike (and, presumably, high levels of unemployment).

All this aside, ironically for the Kirby thesis, Henley suggests that the higher mark-up, for a period of generally depressed prices, arises out of the cost reduction brought about through mechanization. He even considers mechanization to be still more attractive under a cartel since the potential profits of cost reduction are that much greater. Presumably the same applies to amalgamations also as a source of cost decrease. So the source of cost reductions for Henley which allows the mark-up to increase through the cartel is precisely the change that Kirby seeks to argue is obstructed by the cartel! Thus, whatever the validity of Henley's results, they offer no support to Kirby's views.

In short, it is far from clear that Kirby's hypothesis can even be tested in the light of the weakness of the cartels at all times and, to the extent

The coal question

that it can be tested, it appears to be refuted. To the extent that mine owners supported cartel schemes collectively, whilst attempting to evade and avoid them individually, this is probably more a consequence of the weakness of the industry than its underlying cause. Especially where weakness to amalgamate is concerned, the causes have to be sought elsewhere.

Part two
Coal royalties

Chapter three

Royalty or rent: what's in a name?

The purpose of this chapter[1] is to examine a debate that took place amongst economists at the turn of the century on the question of whether a mineral royalty constitutes a rent or not. The first section begins by locating the debate against the background of the development of the British coal industry at that time. Why should the issue have arisen at all? It is argued that the system of landed property in Britain increasingly began to impede the progress of the industry towards the end of the nineteenth century. As the payments of royalties by mine owners for the right to extract coal was the most immediate symbol of the economic influence of landed property, it is hardly surprising that the nature of royalties should have come under the scrutiny of economists.

However, as royalties were only in the first stages of playing an obstructive role, the practical significance of the debate over royalties was not at that stage of any major importance. Indeed, as soon as royalties became a more serious impediment to the coal industry they were taken into state ownership, in 1938, and theoretical as opposed to practical arguments were notable for their absence. Accordingly, around the turn of the century, there existed a momentarily unique opportunity for economists. A conceptual issue was provided for them by the royalty question, but nothing at that time turned on the answer provided. Consequently, the interest stimulated by the royalty/rent distinction provided a terrain upon which other theoretical issues could be engaged that were then more relevant to and internal to the current development of economic thought. As it were, the emerging practical problems provided the grounds for the debate, but the debate itself had entirely different concerns.

This is the focus of the second section where reference is made to the more significant debate within economic theory over the relative merits of partial and general equilibrium analysis. The royalties controversy was a weapon in this second broader debate, although the royalty debate was confined to those who subscribed to partial equilibrium

The coal question

analysis. The competing views are examined in the third section where the different positions held are shown to correspond to the differing views over partial and general equilibrium analysis. Finally, in the last section, some more general conclusions are drawn concerning the relevance of the debate for present-day economics.

The historical background

The source of the interest in the economics of royalties was clearly derived from the British coal industry even if the debate itself made little reference to Britain or to the specific mineral coal. The British industry had experienced an impressive expansion over the fifty years preceding the first world war. From 1850 to 1913, employment increased from 250,000 to over a million. Output increased from 60 million to almost 300 million tons per annum, and exports had grown even faster to absorb almost a third of this total.

Nevertheless, the industry was not without difficulties. It was characterized by pronounced cyclical movements around its trend of expansion. From the last decades of the nineteenth century it began to suffer from stagnation and even declines in productivity which explains why output growth required such a huge workforce.[2] From as early as 1865 Jevons, one of the vanguard in the marginalist revolution, had in *The Coal Question* anticipated this decline. He had visions of coal replacing agriculture as the source of the Ricardian extensive margin through the exchange of coal or coal-based manufactures for food imports.[3] Even so, this merely displaced the inevitability of diminishing returns on agricultural land on to decreasingly fertile mines, ultimately leading to a stationary state, even if temporarily postponed. Finally, a more immediate problem was posed by the rising competition from the development of coal industries in other countries. This entailed competition not only in export markets but also for other exports in which there was a high indirect coal content (such as iron and steel) for which the domestic price of coal was important.

In the climate generated by the Great Depression of 1873, concern for the British economy became expressed in the setting up of Royal Commissions. One of these, the Royal Commission on Mining Royalties of 1890, is of central interest.[4] It had the task of investigating the extent to which the separation of the ownership of land and its minerals from the ownership of the working capital of the mines impeded the progress of the industry and, in particular, whether high coal (and the less significant iron ore) royalties disadvantaged the (competitivity of the) industry.

The Commission set about the issue in two different ways. The first involved an empirical examination of how the royalty system worked and it was carried out in great detail. The Commission discovered that there were a variety and a combination of ways in which royalties were paid (acreage, tonnage, rental) with fixed, dead, or certain rents and sliding scales according to the price of coal. There were also varying lengths and conditions of leases.

Thus, at the immediate empirical level, the idea of a royalty as a payment per unit of coal extracted is something of a misnomer even if this can always be worked out after the event by simply dividing total payments to landlords by total tonnage extracted. Unfortunately, statistics on these payments have always been presented in this way giving a false impression of the simplicity and uniformity of the royalty system.

However, despite the complexities involved, no major objections were made against the royalty system by mine owners. Nevertheless, difficulties were recognized for transporting coal, as this might involve, around the vicinity of the mine, the need to pay a wayleave royalty for the right to carry coal through land that had already been worked out. And there was the need to gain access to railway lines for carrying coal to market and this could generate a rent on neighbouring properties that had no other continuing connection with the mine. Otherwise, there were perhaps no more complaints than might be expected in the negotiation of complex leases.

The second method employed by the Commission to investigate what it termed the 'economic operation of the royalty system' was theoretical rather than empirical and cursory rather than detailed. As mentioned, concern was focused on the effect that royalties had on the price of coal and the manufactures produced with it and, consequently, on the level of foreign competitiveness. The conclusions can be summarized as follows. There is a difference between a royalty and a rent because one involves the removal of a mineral while the other, in principle, leaves the land unchanged. The level of royalties, however, like a rent, evens out the profitability differences between mining and other conditions, such as location.[5] Consequently, the only addition that royalties could make to price is given by the level of the minimum royalty paid, other royalties being price determined rather than price determining, reflecting superior conditions from an otherwise externally given reference point.

These points will be taken up in the third section. For the moment it is important to recognize that the economic assumptions on which they are based have not been linked to the earlier mentioned empirical analysis of the royalty system. What the economic theory simply presumes is that capital can flow freely on to the land and between lands so that royalties drop out into the hands of landlords, technically determined by differing mining conditions quite independently of the system of landownership involved.

These remarks are borne out by the Commission's consideration of alternative systems of royalty ownership. Again their empirical analysis was substantial, as a survey was made of the royalty system prevailing in many other countries. What was found without exception in Europe is that the royalties had been taken into state ownership a hundred years or so before. The reasons for this were made clear by the representatives from various countries who were questioned by the Commission or who submitted written evidence. The patterns of landownership were so sub-divided that, for a reasonably sized mine to be established, terms would have to be arranged with many separate landowners.

Thus, 'it is unanimously admitted that the fertile results follow the absolute distinction that exists between surface property and the working right of mines. In a country where property is so minutely sub-divided as it is in France, the reasonable and active working of mines would be impossible on any other system (than state ownership)'. For Germany, 'in many industrial districts of the country the ownership of the surface is so divided that it would be impossible to carry on deep mining under any other principles' and:

> Besides, in many cases a strata of minerals extends underneath the property of several landowners, and it would be almost impossible to work different mines scattered on the larger or smaller plots belonging to different owners, and to this would come the additional difficulty of arriving at an agreement between the owners with regard to the working of mines . . . it has been arranged in Austria–Hungary to make them (minerals) entirely independent from the landowner . . . . This system has specially promoted the establishment of mines.

The same story is also told for Spain, Portugal, Italy, and Luxemburg; landownership is so fragmented that minerals had been taken into state ownership to promote the development of mining. Without this, capital could not flow freely on to and between lands. By contrast, for Great Britain, this problem scarcely seems to have been considered by the Commission who simply observe that, 'where a large mineral field is the property of one individual no difficulty arises in respect of its full development'.

The reason for this complacency is not difficult to discern. In Britain the pattern of landownership was not fragmented; ownership was highly concentrated and this was even more so in the case of coal royalties.[6] Rather than small landowners obstructing mining through the charges that would be made for the small quantities of coal that they owned, it was more a situation of large landowners positively encouraging numbers of mine owners to extract as much coal as possible. This explains in part the occurrence of fixed rents – to be paid annually irrespective of the quantity of coal removed but against which the royalty charge for the first batch of output would often be set.

This pattern of large landed property made nationalization of the royalties unnecessary in Britain. The same is also true of those countries such as India, Australia, and the Americas, where English law was liable to prevail but in which the coincidence of land and mine ownership had been created. In other words, it was not the state or private ownership of royalties as such that was important as the extent to which large enough coal holdings could be leased to form mines of sufficient size.

Indeed, in many of the European countries, mineral rights were no sooner taken into state ownership than they were sold as concessions to private individuals. These concessions were then themselves subject to trade, possibly in smaller parcels, possibly to be amalgamated. Further regulation of this trade to prevent 'absentee' concessionaries from obstructing the development of mining does not seem to have been extensively necessary. Where it was, state regulation was still possible and this could also be used to guard against monopoly of supply.

Thus, 'the tendency in Belgium is towards the amalgamation of several neighbouring concessions, these being often of a small extent' and:

> It is a common practice for concessionaires to sell or let their concessions to companies who undertake the working of the mines . . . .
> There is absolutely no guard against companies enlarging their holdings through purchase of other undertakings: in fact this is proceeding very rapidly in all the German coal fields without check.

'In the north of France at least, concessions are commonly united and are generally worked by companies' and 'in Austria–Hungary, concessions can be and are freely sold.'

In the light of this evidence, the Commission was perhaps justified in concluding on empirical grounds that the system of private royalties had presented no substantial impediment to the development of the British coal industry. Whilst there was a sharp contrast between state and private ownership of the royalties and in the separation between their ownership and the ownership of the corresponding land surface, different countries in their own ways had provided for sufficiently large holdings of coal to allow mining to take place.

However, the Commission's burden of proof of the acceptability of the British royalty system rested more on theoretical than empirical grounds. It assessed the effects of alternative systems of royalty ownership. If it took them into state ownership then, apart from the minimum royalty, the state would simply charge the same as the private landowners and nothing would be changed. It was also felt that should royalties be abolished, production would be concentrated on better mines and this would unduly dislocate the current prospects of the worst mines by lowering prices.

This argument seems to depend upon a substitution of average for marginal cost of production in the determination of price once royalties are abolished. Whatever the merits of this analytically, the Commission's conclusions are drawn, paradoxically, by refusing to examine seriously alternative systems of landed property. For, in case of abolition/nationalization of the royalties, they presume that the private system is simply reproduced so that whatever happened before is more or less reproduced with the exception that the state now collects the royalties (minus the minimum) and there is some marginal shift of production towards the better mines (whose average are relatively much lower than their marginal costs of production).

Significantly, the abolition/nationalization of the royalties is treated in this way without any specification of the system by which the right to mine a particular piece of land is determined. As it were, it does not matter since the economic consequences are presumed to be predetermined by underlying technical conditions, again revealing that the possible effects of different systems of landed property are already absent.

Similar conclusions are drawn through another line of argument. Suppose a royalty and mine owner coincided. Then, the royalty would accrue to the mine owner, representing an individual abolition of the royalty system. Otherwise, were royalties to be reduced by an individual landowner, then the reduction would simply transfer to the profits of the lessee. A general reduction of royalties across the board (without specifying how this was to be done) would ultimately have the same result, only the consumer might also benefit through the whittling away of excess profits by competition and a reduction in the price of coal.

Again, this can be seen as considering a revision of the royalty system merely in terms of the limited effects of passing on the revenue involved either to the mine owners or to the consumers (as previously to the state where the royalties were nationalized). As state ownership would either require appropriation of private property or payment of compensation, there can have seemed to be little motive to change the system. Otherwise a general reduction in royalties by decree or taxation, even to the limit of abolition, must have appeared as an undue assault on a particular form of (landed) property.

The Commission's deliberations can be summarized as follows. On the basis of their empirical investigations, they came to the view that the royalty system in Britain posed no major problems for the development of the coal industry because capital could flow easily on to coal lands.[7] These empirical observations were then elevated to a theoretical truism: as long as capital could flow freely on to the land, any royalty system would have little or no distorting effect upon the industry. The system of royalty ownership and the mobility of capital are treated independently with the latter being determinant. The size and effects of royalties are separated from the system of landownership, with the latter simply determining the beneficiaries of the royalty revenues – which are themselves determined by the mobility of capital across different mining conditions.

The opinion of the Commission in 1890 on the royalties was perhaps the last that could be so favourable. Mine sizes were expanding fast, partly in pursuit of seams within existing mines, and partly as deeper and more difficult working conditions demanded more coal to be extracted across a wider area to spread the fixed costs involved. Prophetic evidence for this is to be found in the Commission's investigations, as mining became increasingly dependent on relations between more than one landowner, initially in arranging transport and wayleaves rather than multiple leases. These were to follow. As mine size expanded relative to a given pattern and distribution of landownership, so the same obstacle that had plagued the European industry in its infancy, and necessitated state ownership of the royalties, had come to inflict the British industry in its adolescence.

Not surprisingly, the associated difficulties, once confronted, matured extremely rapidly – so much so that by 1919, the Sankey Report found mine owners' representatives unanimously supporting nationalization of the royalties (whilst, unsurprisingly, equally unanimously opposing nationalization of the mines). The reasons for their leap into such a policy, and its resonances with socialism, are best found reported elsewhere. The Acquisition and Valuation of Land Committee of the Ministry of Reconstruction, the Scott Report (1919), discovered an intensification of the problems investigated by the Royal Commission of the 1890s, together with the difficulties of extending mines across the boundaries of surface ownership.

By 1925, the Samuel Report had found that on average each mine required five leases and reorganization of the industry depended on nationalization of the royalties. In the meantime, in a wave of legislation in the early 1920s, the rights of private landowners were reduced and those of mine owners extended by the granting of certain compulsory powers. This, as well as the fear of royalty ownership as the thin end of the wedge for nationalization of the mines themselves, temporarily reversed the mine owners' commitment to public ownership of royalties.

As a result, nationalization of the royalties was delayed until 1938. In the next and some later chapters, it will be argued how this could have impeded not only the rational organization of the industry in mine layout within and across the patterns of landownership, but also have obstructed the mechanization of production. This is not, however, strictly relevant here. For the Mineral Commission of the 1890s separated empirical or institutional conditions in which the royalty system operated from the theory of royalty determination. This analytical tradition was carried over into the interwar years, a period when the royalty system was considered problematic. Consequently, the dispute over royalty nationalization became practical rather than academic: did the system operate smoothly or not? For this, the conceptual distinction between royalty and rent proved a rarified luxury confined to the ivory tower.

This allows the timing of the debate between economists over the distinction between royalty and rent to be explained. Towards the end of the nineteenth century, the private system of royalty ownership in Britain was beginning to produce problems in organizing the industry but these were not severe enough to prevent a complacent attitude being adopted towards them. Consequently, debate over the royalty system did not concern the practical organization of mining in relation to landed property. Instead, controversy could be elevated to a theoretical plane independent of the system of landed property, as has been seen for the Minerals Commission.

Here, by way of comparison, there is a striking analogy with Marx's (1969) analysis of Ricardo's theory of rent. Marx suggests that the assumptions behind Ricardo's theory make it quite independent from the type of landed property under consideration and this is a (partial) reflection of the conditions of British agriculture which must have seemed correspondingly peculiar to those European economists confronted by Ricardo's economics:

> Both of them (Ricardo and Anderson), however, start out from the viewpoint which, on the continent, seems so strange: 1. that there is no landed property to shackle any desired investment of capital in land; 2. that expansion takes place from better to worse . . . 3. that a sufficient amount of capital is always available for investment in agriculture.
>
> (p. 237)

Something similar happened in the case of the debate over royalties and rent. Assumptions such as Ricardo's were made and must, by the interwar period, have seemed bizarre, even to those in Britain, confronting the problems posed by landed property in the development of the coal industry – that is if they ever bothered with the theory rather than the practical matter to hand. Thus, while the debate over royalty and rent was produced or stimulated by a specific conjuncture in the development of the British coal industry – just when the royalty system was producing problems that were not too severe – the debate itself had little or nothing to contribute to the solving of the problems of which it was the offspring even when these problems became more severe. How could it do so when reform of the system of landed property lay outside the scope of the theory in so far as any such reform was considered to have negligible effects.

However, despite or, more exactly, because of the practical irrelevance of the debate, its content and the form that it took could be determined by other matters. These were internal to the development of economic theory at the time as will be revealed in the next section.

The theoretical background

In the previous section it has been shown that interest in the economics of royalties arose in conditions where the practical problems of the royalty system were considered negligible. Consequently, the debate over the royalties could pose its own theoretical problems and turn a blind eye to the system of royalty ownership. Conversely, during the interwar period, when the royalties increasingly became a burden on the industry, the problems were perceived to be practical and concerned with the intervention into the system of landownership for which the theoretical assumptions of the free flow of capital were irrelevant along with the royalty/rent debate.

In short, the royalty/rent debate is located around the turn of the century and, rather than being concerned with mining as such, it was used as an instrument in a debate within economic theory. Before considering the debate itself, it is first essential to uncover what the debate was really about by examining the relevant economic theory of the time within which it took place.

For those economists trained in the modern concepts and techniques of mathematics, the passage from the marginalist revolution of the 1870s to the present day might be seen as the uncontroversial evolution of the perfections of general equilibrium theory. This is not what happened. While economists such as Walras and Jevons had an idea of the simultaneity involved in general equilibrium determination, the principles of marginalism after the 1870s were initially applied within a partial equilibrium framework. This is most notable in the work of Alfred Marshall. At the same time, economic theory was informed by the principles of general equilibrium theory but in an uncertain way that lacked the confidence of today's practitioners.[8]

The reason for this is not to be found exclusively nor necessarily predominantly in the newness or difficulty of the mode of thinking – in the use of mathematical techniques, for example. But, there was a conceptual problem involved which made economists at the time hesitate from embracing general equilibrium theory even though the logic of their analysis inevitably drove them into its arms. One major result of general equilibrium theory is to eliminate the causative significance between different factor inputs as the sources of revenues. For revenues become something of a misnomer since they are simply derived from factor prices, and all of these prices are determined simultaneously and by exactly the same marginal principles regardless of the input to which they apply, whether it be land or labour, for example.

Consequently, distinctions made at the level of revenues such as wages, profits, and rents can only be maintained by distinctions drawn over the conditions of supply and demand which are specific to labour, capital, and land. Rent, for example, is then explained in terms of land in fixed and indestructible supply. A royalty might then be reserved as the term to characterize the remuneration due to a factor in fixed but destructible supply. It becomes simply a question of names and not principles of determination that mark the differences between one source of revenue and another, for the analytical principles remain invariant whether rent, royalty, wages, or profits are under scrutiny. Every factor of production has its own price determined simultaneously with that of other factors by the conditions of marginal productivity and utility. Thus, distribution involving wages, rent, and profit is simply a corollary of price theory and has no specific conceptual nor causal status of its own.

Some economists were unhappy about the conceptual loss involved in moving to general equilibrium. As this was itself based on generalizing, through marginalism, the Ricardian principles of rent determination to the economy as a whole, it is not surprising that the problem was felt most acutely within rent theory.[9] It gave rise to a debate over whether rent is price determined or not.[10] Now, for general equilibrium theory, such a debate is ridiculous since all prices, including that of land, are determined simultaneously by an interdependent system of supply and demand across many sectors (as determined by production possibilities and individual preferences).

But it is precisely this which eliminates the specificity of land as a source of revenue (apart from its hypothesized fixity of supply) as is made clear by Jevons' maxim, 'so far as costs of production regulates the values of commodities, wages enter into the calculation on exactly the same footing as rent'. Logically, those who shied away from these implications of general equilibrium could do so within the marginalist framework only by using partial equilibrium analysis. In a one-good world, rent would be price determined according to the differential productivity of better over the marginal (no rent) land in use. Moreover, a particularly distinct role could be assigned to land as being a source of differential productivity and hence it appears as a separate sort of factor input and revenue source, as in Ricardian rent theory.

Because the conceptual specificity of rent required partial equilibrium analysis within the marginalist school, the debate over rent theory was a debate between the relative merits of partial and general equilibrium and to that extent a 'dialogue of the deaf' (Wessel 1967).[11] Each antagonist could be right about a particular concept of rent and its relation to land. The debate was, however, complicated by the different types of partial equilibrium that were used. One might assume a one-good world while another might assume many goods but for which all prices were fixed except in a single market under consideration. As economic models, there is little to choose between these partial analyses since other prices will enter as exogenous technical constraints. But conceptually the existence of other goods with a price is 'closer' to general equilibrium than is a one-good model.

This section can now be brought to a close by bringing out its significance for the debate between a royalty and a rent. It follows immediately that the debate can only have been of interest to those who wished to distinguish conceptually between different types of revenue – for a general equilibrium theorist would have no interest in the matter, just as there would be no interest in distinguishing between wages, profit, and rent, let alone between sub-categories of the latter. Accordingly, the debate over the royalty/rent distinction can only be conducted within the confines of partial equilibrium. Those who distinguish rent from other factor incomes through the use of partial equilibrium implicitly recognize, in contrast to general equilibrium theory, that the access of capital to land differs from its access to industry more generally.

Pursuing this one stage further, those who distinguish a royalty from a rent recognize that access of capital to mining or extraction differs from its access to land for other (agricultural) purposes. But it only makes sense to have a debate over this with those who are at least willing to accept that rent is distinguishable from wages and profits, that is with those working within a partial equilibrium in so far as analysis is confined to the neoclassical school.

The debate

The debate over whether a royalty is a rent or not did not concern the issue of what gets to be called a royalty; this was generally accepted to be the payment for the right to remove a mineral only available in fixed supply. Rather the debate was concerned with whether or not a royalty was caused in the same way as a rent, and this is why it is an irrelevant debate for the simultaneity of general equilibrium theory which is unable to assign a unique causative significance to factor inputs and their associated revenues.

It is as well to begin a review of the debate with Ricardo since, in his theory, rent is determined in a manner distinct from that of wages and profits, and he stumbled upon the future debate in content if not in name. Consequently, the embryo of the later dispute between those who did and those who did not identify a royalty with a rent is to be found in Ricardo (1971). He is concerned with the principles of determination and not simply with the names of the various factor incomes. His desired solution is to determine the rent of mines in exactly the same way as the rent of (farm) land (pp. 108-9). This is done by reference to the differential productivities of the original and indestructible properties of the land. But because Ricardo's theory depends upon indestructibility, it is inappropriate for mineral extraction – a problem that he appears to have neglected. He does refer to timber removal but in the context of the timber itself having a value determined by its costs of reproduction (pp. 91-2). Later, he also removes the condition of original powers but in order to allow improvements, no matter what their origin, to be incorporated into the indestructible properties (p. 268). What is clear is that Ricardo's rent theory raises the problem of destructible conditions (such as the presence of minerals) only to leave it essentially unconsidered.

Nevertheless, Ricardo's approach can be judged to suggest two solutions to the pattern of the rents of mines, each solution requiring a partial equilibrium framework. Mines may be treated as land in general, as if they satisfied indestructibility and as if there is a one-good world in which rents are price determined, equalizing the residual between costs on better and worse mines in use. Alternatively, when destructibility is recognized, the value of the minerals is predetermined and enters into the price of the extracted commodity as a royalty, distinct from the rent that is price determined and reflects the differential fertility of the first case. Ricardo himself, however, only predetermines the value of the mineral (in this case timber) by allowing it to be reproduced with an associated value so that indestructibility is restored.

The property that these solutions share in common is that they depend upon partial equilibrium analysis. For the first solution, the question of whether there is a distinction between the royalty of a mine and the rent of a land must be answered in the negative for a one-good world since rent simply captures differences in costs of production. For the second solution, there remains the question of predetermining the value of the mineral, a problem that ultimately creates circularity in the simultaneity of general equilibrium since the mineral's value will affect the cost of its own extraction as well as depending upon final levels of demand.

The debate over rent and royalty, for which Ricardo was the precursor, led to the adoption by protagonists of one or other of the solutions outlined above. The view of the Royal Commission on Minerals, outlined earlier, can now be reassessed in the following terms. A royalty is distinguished from a rent on the basis of the distinction between destructibility and indestructibility. Consequently, a royalty is made up of two parts, one reflecting mining conditions and corresponding to the normal Ricardian notion of rent determination, and the other part corresponding to a minimum royalty which reflects the value of the mineral extracted. As it were, if the mineral were simply easily available, without extraction costs, it would have a scarcity value making up the royalty component.

Sorley (1889), Orchard (1922), Flux (1923), and Marshall (1959) supported this latter position that a royalty was distinct from a rent. For Sorley, who appears to have been the brains behind the Commission's thinking, a minimum royalty entered the price of coal as the price of the mineral to be extracted together with a compensation for the loss of beauty to the land. Here is seen the demand considerations associated with utility entering quite openly as causative factors (and as a prior charge). Marshall argues that:

> A royalty is *not* a rent, though often so called. For, except when mines, quarries, etc., are practically inexhaustible, the excess of their income over their direct outgoings has to be regarded, in part at least, as the price got by the sale of stored-up goods – stored up by nature indeed, but now treated as private property; and therefore, the marginal supply price of minerals includes a royalty in addition to the marginal expenses of working the mine . . . the royalty itself on a ton of coal, when accurately adjusted, represents the diminution in the value of the mine, regarded as a source of wealth in the future, which is caused by taking the ton out of nature's storehouse.
>
> (p. 364)

Thus, for Marshall, the royalty represented the price of the mineral in the ground to which the expenses of working the marginal mine had to be added to determine the (uniform) price of the extracted mineral.[12]

The preceding authors are relatively close to general equilibrium theory because they rely upon a partial equilibrium in which there is another good, the unextracted mineral, even if its price is predetermined. Consequently, to use Marshall's terminology, the remuneration to the landowner includes a 'producer surplus' for differences in extraction costs, plus a royalty determined exogenously by supply of and demand for the extracted mineral.[13] The royalty is itself not caused in principle by anything which distinguishes it from any other price or revenue, it is merely distinguished by being fixed in supply and destructible.

In short, this group sees the market for coal as establishing a price which can be conceptually divided into two parts, one for the scarcity value of the mineral in the ground, and one for the costs of extraction. As these costs will differ according to relative mine fertility, the differential costs will make up the rents of the land, to be distinguished from the royalties. This is of necessity a partial equilibrium analysis since the interrelations between the rest of the economy and the coal sector and between the scarcity value (royalty) and costs of extraction are assumed away. The methods of extraction, for example, may themselves affect the demand for coal, directly in the generation of power or indirectly in the demand for machinery which takes coal to be produced.

This is the point of departure for a competing group of economists. These do not see how the value of the unextracted mineral can be calculated (as the royalty) independently of the conditions of extraction. In contrast to the distinction between royalty and rent, Gray (1914) and Taussig (1939) argue that the two are inseparable. Taussig in his rent theory adopts a one-good world, 'rent forms no part of the expenses of production; that is, it forms no part of those expenses of production which affect price' (p. 96). This thinking is carried over into the consideration of mines for which the value of an unextracted mineral and its costs of extraction are indistinguishable. For him, the last mine in use will pay neither rent nor royalty, at least in theory, since it is on the margin of use and so the existence of an independent royalty distinct from a rent is denied (p. 140). Gray essentially puts forward the same argument, adding that only in accounting terms can the loss in the value of a mine due to the extracted mineral be attributed to a royalty.[14]

For these authors, there can be no value of the mineral independent of the costs of extraction and so, in a sense, they rely upon general equilibrium considerations. On the other hand, they do so in a partial equilibrium in which rent/royalty can be reduced to differences in costs of extraction, essentially a one-good world in which the exhaustibility of the resource is secondary to the costs of extraction. Taussig, for example, specifically denies the difference between royalty and rent on the basis of sand and clay being available in abundant quantities and not commanding a royalty despite being exhaustible. Thus, to put it another way, in contrast to Marshall *et al.*, only producer surplus is permitted to determine the payment for the right to mine and no allowance is made for the consumer surplus implicit in the value of the mineral extracted for which a royalty might be defined.

Concluding remarks

To summarize: the royalty/rent debate was stimulated by the increasing but temporarily moderate problems associated with the British royalty system. As the problems subsequently intensified, they were ultimately resolved, as elsewhere in Europe more than a century before, by the practical measure of the nationalization of the royalties at the end of the interwar period – a measure for which theoretical support and justification seemed unnecessary.

This sequence determined the timing of the debate but not its form nor its content. This depended more upon developments internal to economic theory and which may be considered equally chronologically contingent. The hegemonic thrust towards general equilibrium was initially resisted by those understandably reluctant to abandon distinct distributional theories of the separate class revenues – wages, profits, and rents. Thus was created a place for a minor sub-plot in terms of whether different revenues from within the income provided by land could be distinguished.

Debate could only take place within partial equilibrium, for otherwise there was no distinction between let alone within revenues. Consequently, the debate gives rise to something of a paradox. For those closer to general equilibrium theory, in the sense of having a multi-good model but with all prices fixed except for one, were able to distinguish a royalty from a rent by having the royalty composed of a predetermined scarcity value which would supplement the rent associated with differential costs of extraction. By contrast, those furthest from general equilibrium, with a one-good world of extraction, were unwilling to distinguish a royalty from a rent, for scarcity value and extraction cost were jointly and not separately determined.

But what are the implications of this debate which even at the time must have appeared somewhat esoteric, if not even more so with the benefits of hindsight? In chapter 4, some lessons for the significance of the debate for positively understanding the role of landed property will be addressed. Here, two lessons for understanding economic theory will be presented.

First, it is worth recognizing that the confusion over royalty and rent remains although this is not made explicit because of the casual method in which the concepts are employed. Thus, Dasgupta and Heal (1979) offer a book which is exemplary in setting out in mathematical terms the propositions associated with the economics of exhaustible resources. In doing so, it makes clear the extent to which little progress has been made in understanding the role of landed property in economic development. The same confusions over royalty and rent are reproduced in so far as a royalty is seen 'as the competitive value of a pool of oil or a deposit of coal' (p. 159).[15] But is this an accounting identity, as for Gray, or a condition of access to the land as for Marshall – price determined or price determining?

An answer can be given in terms of the partial equilibrium model employed. But these models can tell us nothing about the actual conditions confronting capital as it seeks to gain access to the land. It tells us only about the properties of the model; indeed of the properties of an economics that is remarkably unsuitable for distinguishing the significance of different types of property relations – hardly surprisingly as it seeks to deny this significance and reduce all such relations to a single dimension of payment.

To some extent this anticipates the conclusions of the following chapter. A second more general observation to make about economic theory concerns the different influences by which it progresses even if within the narrowly defined terms laid down by the marginalist revolution. Certainly, economic theory is subject to external stimulus as economic problems are thrown up, in this case the royalties. However, this does not determine the theoretical response which may, as has been shown, be orthogonal to the practicalities involved. Rather, the external stimulus may simply be incorporated into the internal logic of the present orthodoxy.

An apt confirmation of this is provided by the more recent revival of interest into the economics of exhaustible resources, as argued by Fine and Murfin (1984). They point to the flagging literature of optimal growth theory which was given a new lease of life in the early 1970s by the economics of exhaustible resources. Whilst the questions posed were no doubt stimulated by the oil crisis, the answers given have been far removed from the realities of raw material supply, with competitive or monopoly conditions of supply extended into the indefinite future through the use of optimality techniques.

In this, there is a striking parallel with the debate over royalty and rent, since this took on a life and purpose of its own, removed from the problems of the British coal industry. At least, in the latter case, there was a dispute reflecting the uneasy passage from classical to neoclassical economics. Today, on offer are pages of equations whose level of formal validity and rigour stands in sharp contrast to their level of relevance.

But finally, a direct response must be given to the issue of whether a royalty is a rent or not. The answer is both yes and no! Yes, in so far as each represents the revenue that accrues to the landlord in return for capital's access to the land for mining or farming. But also no, in so far as those conditions of access are different and must be specified in each case. This chapter has revealed the heroic attempts of some marginalist economists to specify a distinct royalty by positing a predetermined value of the mineral in the ground. The irony is that this could never be entirely satisfactory within the neoclassical approach since this seeks to reduce conditions of access to the land or, in other words, social/class relations of production on the land, to a matter of inputs and their costs. For this the unsatisfactory logical conclusion for neoclassicals in all but partial equilibrium has to be that royalties and rents are the same (but much the same is also true of wages and profits).

Chapter four

Royalties: from private obstacle to public burden?

Chapter 3 has examined how economists viewed the royalty question at a particular conjuncture. This chapter is firstly concerned with the views of the trade union representatives of the miners, particularly as these were presented to the Royal Commission on Minerals in the 1890s.[1] Whilst they unsurprisingly adopted a much more critical approach to the issue, nonetheless, it will be shown in the first section that the analytical content of their opposition remained quite limited. This probably had little practical effect in limiting the scope and direction of trade union activity, for there were more prominent conflicts at issue concerning wages, health and safety, and even nationalization of the mines. Also, miners did tend to support nationalization of the royalties. However, it is worth examining their arguments because they are typical of those presented in radical opposition to landed property in a wide variety of circumstances. Accordingly, a more general critique of radical approaches to landed property can be obtained by revealing the limitations of the miners' views. The implications of this are taken up in the final section.

Before that, there is a discussion of the role of the royalty system, particularly for the interwar period. A dialogue is run between the idea that the fragmented royalties posed a practical problem for organizing and reorganizing the industry and the idea, as for the miners, that they were an economic drain on the industry in the revenue absorbed. The conclusion drawn is that this dichotomy is a false one but, to combine the practical with the economic, requires a more sophisticated theory of landed property, sensitive to the material circumstances of the industry. In the concluding remarks it is suggested that Marx's theory of rent is well suited to this purpose particularly as it is directed towards the dynamics of capital accumulation in historically specific conditions.

The miners' attitudes to royalties

In the latter half of the nineteenth century there was much agitation over the unequal ownership of land in Britain and the demand for it to be nationalized was raised. Indeed, the socialist demand for the public ownership of the means of production often takes land as its first target even if, especially in the UK, this demand has fallen out of the limelight. (See Eldon Barry 1965.) A survey, undertaken in the 1870s and intended to reveal how widely dispersed landownership had become, had the embarrassing effect of proving exactly the opposite.[2] Subsequently, no other comprehensive survey has been attempted. Against this background, it is hardly surprising that the trade union movement should take up the demand for the nationalization of the wealth underneath the ground as well as for its surface use.

The Miners National Conference in 1889, for example, welcomed the setting up of the Royal Commission on Mining Royalties and expressed their hope 'that the outcome of its labour may be the full and complete restoration of the mineral to the State'.[3] The first annual conference of the Miners Federation of Great Britain (MFGB) brought together the separate and fiercely independent regional unions representing workers from the different coal districts. It raised on its agenda 'Mining Royalty Rents: what position shall this Federation take with regard to the Commission in getting evidence, etc?' At the TUC of 1891, a motion was passed supporting the nationalization of the minerals and mines, calling for a 'Bill for restoring to the country its property in minerals and metals in terms of statute laws'.

Despite this apparent unity of feeling, there were divergencies of opinion amongst the miners' representatives. To place this in perspective it is necessary to be aware that these differences cut much deeper and extended much more widely than the narrow issue of royalties. For they centred on much more immediate matters. On the eight-hour day, for example, the districts of the North East were opposed to legislation, and this entailed their remaining outside the MFGB. Politically, there was a strong tradition of Liberalism amongst many of the miners and their leaders; this expressed itself in a conciliatory attitude to the relations between capital and labour and a wish to work to the mutual benefit of both – an approach associated with respect for property rights and presumably extending to landed property.[4] Certainly, at the time of the Royal Commission on Mining Royalties, there was no unity of opinion as is seen by the evidence presented by miners' representatives in answering questions from the Commissioners.

The diversity of opinion amongst the miners is paradoxically indicated by the futile attempts of the Commission to get Ben Pickard to attend as a witness. Pickard, President of the MFGB, had been 'out of town' when first invited by letter to appear. He then declined to give evidence on the grounds that he could add nothing beyond what had already been offered by other unionists, Cowey, Haslam, Woods, and Aspinwall, who had acted as witnesses in the meantime. His unwillingness to appear may have reflected a wish to avoid adjudicating between the various opinions of his colleagues, so delicate were the ties holding the MFGB together. Alternatively, he may have thought the whole exercise a waste of time to pursue any further. For, it is possible to read the relevant Minutes of Evidence as a patronizing exercise on the part of the Commissioners, in which they indulged themselves in teaching the miners a few lessons in the elementary principles of economics and private property.

In terms of the effects of the royalty system, there was amongst the miners some, but very little, disagreement. Their evidence as a whole is best considered in terms of a divergence of experience rather than of opinion, although there was also some uniformity of experience.[5] Stoppages of production due to excessive royalties or intransigent landlords were rare, but not unknown, and were in any case difficult to prove not to be due to other factors.[6] Many of the miners, however, commented upon a distributional conflict between royalties and wages, observing in particular that the mine owners often argued that they were unable to afford higher wages because of the high level of the royalties.[7] This proved a source of industrial conflict between capital and labour over sums that were quite often lower than those paid for the royalties.

A more sophisticated relationship between wages and royalties was also suggested.[8] The payment of royalties meant that less capital advances were available to pay wages, even if labour could be profitably employed, since the royalties were a first call upon the finance of the mine owner. This was acutely felt in the extreme circumstances when a mine owner was bankrupted so that even the wages of work already done might be lost.

In terms of foreign competition, the miners presented the case that royalties raised the price of coal and this led to the loss of export markets.[9] With experience of this, the Commissioners learnt to prepare a trap. Asking whether this view was based on the belief that royalties were higher in Britain than abroad, the miners, on answering in the affirmative, were implored to inform their colleagues that this was not so and that they should abandon any discontent with the private ownership of the royalties. There was also some recognition by the miners of the practical difficulties arising out of the private royalty system. These concerned coal barriers, wayleaves and the need for collective pumping of water (see the next section and chapter 5). Not surprisingly, these could be more extensively documented by engineers and by the mine owners.

Where there was less agreement amongst the miners was in what to do about the royalty system. Not all agreed that state ownership was desirable although this was the majority opinion.[10] Those in favour of state ownership were of the view that this would directly or indirectly benefit the miners' real wage, although Smillie also saw the state as using its powers as leaseholder to guarantee provisions for safety. The Commissioners, however, were keen to take the argument one stage further. If, they argued, the miners or anybody else were to benefit from the state ownership of the royalties, it could only be at the distributional expense of the current royalty owners according to the extent that they were paid compensation. This led to a discussion of the level of compensation to be paid, with 'harder' or 'softer' lines being adopted by the miners in correspondence to their ethical judgement of the worthiness of entitlement to produce of the land.[11] Keir Hardie, for example, merely wished to compensate widows and their children whilst Woods argued for full compensation to be paid to all.

To a great extent, these discussions were conducted on the basis of the royalties being determined geologically as a differential rent and, otherwise, as a payment for the mineral. Each of these takes the royalty to be independent of the system of landed property and so all that is at stake is the identity of the recipient of the income generated. The same approach is characteristic of discussion of proposals to abolish royalty payments, whether through state ownership or not. In effect, abolition of royalties as a whole was seen as the generalization of an individual royalty owner foregoing payment from the mine owner. This allowed abolition of the royalties to be discussed, as if it were simply the reduction of a cost, without the necessity of proposing an alternative method of determining who would have access to the land for the purposes of mining and under what conditions.

Thus, the effect of abolition could be considered in purely distributional terms according to whether the lost royalty were recovered in the profits of the mine owner, the wages of the miner, or in a reduced price of coal. This led the miners to accept that, although they might be reduced, the royalties could not be abolished without creating unequal rewards within the industry. Those that refused to recognize this were dismissed by the Commissioners, following a patient lesson in Ricardian political economy on the role of differential rent, supplemented by the suggestion that the buying and selling of mineral rights might be best left to the self-interest of the private parties concerned.

When examining the royalties analytically in the context of either their abolition or state ownership, both the Commissioners and the miners tended to see them either as a naturally arising revenue to be (re)distributed according to their fancy or as a production cost to the system as a whole in parallel with the cost to an individual mine. Thus, that the royalties were not only a differential revenue but also a differential cost was not entirely neglected although the two economic roles were rarely integrated with each other, particularly as the effects on production were generally assumed away.

However, Young argued that the effect of abolishing the royalties would be to disadvantage those mines that were currently paying the lowest royalties and these would be forced out of business. As he thought this applied to Northumberland whose miners he represented, he opposed abolition of the royalties because of the likely effect on unemployment in his own district. It is not the self-interest of this argument that demands attention so much as the rare focus that it contains on production. Essentially, it recognizes that the royalty system has an effect on where mining takes place and even suggests that it sustains production at the less efficient sites. Paradoxically, the miner presenting this view is the least progressive of the representatives, with those demanding state ownership of royalties on distributional grounds unable to explore the impact of the royalty system on the organization of production.

This situation had not so much changed as become irrelevant by the interwar period. Out of adversity and diversity, the miners had created by the end of the First World War a united and militant union whose campaign over wages, hours, and health and safety increasingly took the form of the demand for nationalization. The argument for nationalizing the royalties as a means to reorganize the industry could be left to be formulated by the liberal bourgeoisie. The miners set themselves the task of mobilizing support by relying more upon distributional arguments. The money paid to the royalty owners could more equitably be used to provide the miners with higher wages, shorter hours, and better working conditions, particularly as the royalty owners often owed their position to the fortuitous acquisition of the coal centuries ago when its presence was not even suspected. Nevertheless, the primary focus of the miners was upon the mines rather than upon the coal, although the nationalization of the first almost certainly entailed the nationalization of the second.

Practical or economic obstacle?

In chapter 3, some emphasis has been given to the way in which the development of the British coal industry should, as mines became larger and pursued seams underground, increasingly spread itself across a number of neighbouring properties. Undoubtedly, this gave rise to a number of practical difficulties, some legal in the negotiation of leases and the interpretation of mining and property law, and some technical in the efficient extraction of coal and the maintenance of the fabric of the mines themselves. In the previous section, it emerges, unsurprisingly, that, whilst the miners were aware of these problems from their own perspective, their concern was primarily with the economic burden that the royalties imposed upon the industry and, through this, on them. This give rise to an emphasis upon distributional factors. Less for the landowners, more for them – the means to be public ownership.

The strength of feeling against the private ownership of royalties is perhaps not surprising. Historically, the landowners had come to own the coal (and other base minerals) in the ground, as opposed to gold and silver, which remained in the hands of the Crown, as a result of the defeat of the Queen in a dispute with the Duke of Northumberland, the decision dating from 1568. As Nef (1932: 318) observes, this led to a peculiar terminology to describe the subsequent exercise of an individual's property right:

> In Great Britain, the meaning of the word (royalty) has undergone, in fact, a curious inversion. An attribute of sovereignty in feudal times, when sovereignty was decentralised, the *regale* has been absorbed, not by the sovereign state, as in France and most other continental countries, but by the landowners. Thus a word originally applied to the rights of the sovereign as against the subject, is now applied to the rights of the subject, as against the sovereign.

There is here an interesting slippage from the landowner to the subject. For the legal right applies to the latter whilst its practical application is restricted to the former, the owners of coalbearing land alone. From this perspective, it must have seemed in the UK that little had changed from the earlier system of decentralized sovereignty. For not only were the royalties extremely concentrated in ownership, the main proprietors were often in direct lineage from the aristocracy of the Elizabethan times and earlier. No doubt this served ideologically to display the royalty system as a feudal anachronism as well as an ethical outrage and economic injustice.

Consider, for example, the findings of the Sankey Commission. From returns to the Inland Revenue in 1918, it reported that £2,939,224 of royalty income was received by the top 100 owners out of a total of £5,960,365, there being 3,789 landowners concerned.[12] At the time of the nationalization of the royalties in 1938, little had changed. The Coal Commission (1945), responsible for paying compensation to landowners upon nationalization of the royalties, reported that over half of the sum that it handed out went to 114 claimants who had received more than £100,000 each. Of the rest, almost 8,000 drew less than £1,000 each, and only 1,300 were paid more than £5,000, there being 13,482 claimants in total.[13] Just to confirm the connection with the regality of the past, the two largest owners were the Ecclesiastical Commissioners for England and the Commissioners of Crown Lands; church and royalty, respectively!

This is the picture across the country as a whole. Within the individual coal districts, concentration was often even more pronounced and

From private obstacle to public burden?

necessarily so given a limited geographical spread of individual ownership. This is illustrated by the tables 4.1–4 for Scotland.[14]

Table 4.1 Distribution of royalty ownership across regions

Districts	No. of claims	No. of non-zero assessments	Total valuation
1	167	116	3,004,839
2	79	69	1,312,331
3	643	459	3,206,334
4	248	171	1,539,281
Scotland	1,137	815	9,062,785

Source: Brunskill et al. (1985)
Notes:
District 1 – Fife, Clackmannan, Kinross, and Sutherland
District 2 – Lothians (Mid and East)
District 3 – Lanarkshire, Linlithgow, Stirling, Renfrew, and Dumbarton
District 4 – Ayrshire, Dumfries, and Argyll

The number of zero assessments is explained by claims for compensation that were rejected, and all other figures reflect claims that had been assessed and possibly modified.

Table 4.2 Concentration of ownership of coal royalties in Scotland

Districts	% Share of top			
	2%	5%	10%	20%
1	37.4	61.1	79.3	92.0
2	22.1	35.2	58.4	80.2
3	41.6	58.3	72.1	85.3
4	40.1	66.5	79.6	89.6

Source: Brunskill et al. (1985)

Table 4.3 Frequency distribution of royalty ownership in Scotland and the UK

Amount of compensation	UK	Scotland districts			
		1	2	3	4
£100,000 and over	1.2	6.9	4.3	1.3	3.5
£ 75,000 and over	1.7	9.5	7.2	1.3	3.5
£ 50,000 and over	2.8	12.1	13.1	2.2	4.7
£ 25,000 and over	5.2	16.4	23.2	5.9	7.0
£ 10,000 and over	9.5	23.3	31.9	13.3	12.0
£ 5,000 and over	13.9	36.2	46.4	22.7	25.1
£ 1,000 and over	27.5	59.5	69.6	53.8	46.2
£ 100 and over	53.1	89.7	87.0	87.8	78.4
£2 and over	100.0	100.0	100.0	100.0	100.0

Source: Brunskill et al. (1985)
Note: All compensations below £2 have been excluded. There were none for Scotland, but 1,565 for the UK. These may have been allowed merely to permit expenses to be paid to claimants.

The coal question

Table 4.4 The top ten royalty owners in the four districts of Scotland

District 1	Name	Amount of compensation received £	Proportion of royalties owned in that district %
1	Earl of Wemyss	504,729	16.80
2	Commissioners of Crown Lands	483,292	16.09
3	Fife Coal Company	406,767	13.54
4	Moray Estates Company	173,374	5.77
5	Earl of Minto	138,897	4.62
6	J. Wilson, Bart	128,813	4.29
7	Rosslyn Trust	126,200	4.20
8	Kerse Estate	118,092	3.93
9	Alloa Coal Company	90,590	3.02
10	Earl of Mar and Kellie	78,219	2.60
District 2			
1	Lothian Coal Company	243,061	18.52
2	P. Dundas, Bart	117,734	8.97
3	Commissioners of Crown Lands	100,846	7.68
4	Buccleuch Estates	92,764	7.09
5	Rosebery Estates	84,487	6,44
6	Earl of Wemyss	67,307	5.13
7	Marquess of Lothian	60,026	4.51
8	J. Don-Wauchope, Bart	57,058	4.55
9	Hopetoun Estate Company	53,642	4.09
10	Penicuik Estates	39,746	3.03
District 3			
1	Hamilton	380,472	11.87
2	Douglas and Angus Estates	219,312	6.84
3	Carron Company	174,111	5.43
4	A. Baillie, Bart	165,829	5.17
5	James Nimmo and Company	112,587	3.51
6	Coltness Iron Company	103,817	3.24
7	William Baird and Company	65,703	2.05
8	Gartshore Trust	61,119	1.91
9	Keir and Cawdor	50,832	1.59
10	C. Forbes	50,143	1.56
District 4			
1	Baird and Dalmellington Ltd.	240,645	15.63
2	Mountjoy	187,319	12.17
3	Buccleuch Estates	143,466	9.32
4	Major J. Oswald	116,137	7,54
5	J. Boswell	106,992	6.95
6	Mrs. C. McAdam	100,011	6.50
7	A. Kenneth & Sons	56,013	3.64
8	Major-General C. Alexander	53,114	2.92
9	R. Angus	45,015	2.70
10	J. Kenneth	41,565	1.86

Source: Brunskill et al. (1985)
Note:
Included in the compensation received by the Earl of Wemyss are those for the Wemyss Colliery, the Wemyss Trust, Captain M. Wemyss. Barry Supple has remarked that the Earl of Wemyss has no connection with the Wemyss Trust and Colliery. We have discovered that Captain Wemyss was responsible for the dispersal of his funds in the Wemyss Trust; although there remains the possibility that the Earl of Wemyss was completely separate from the Trust Fund, his compensation was in the region of £66,000, 2% of the total, making him alone the tenth largest landowner, yet not warranting recalculation of the total proportions: 'The Wemyss Estate were the owners of the land surface in the whole of the area covered by Buckhaven and Methil Town Council, as well as the surrounding area. The Estate also had become the principal owner of the mines. There had been financial jugglery carried through. The Wemyss Estate had sold the mineral rights to the Wemyss Collieries Trust, which was Wemyss Estate's Man Friday. And the Wemyss Collieries Trust sold to the Wemyss Coal Company the right to exploit and develop these mineral deposits. Then, over and above all, for general development the Wemyss Estate had formed the Wemyss Developments Society.' (McArthur, J. (1981) in I. MacDougall (ed) *Militant Miners*, Edinburgh: 153) Included in the compensation received by P. Dundas, Bart are those for the Trustees for R. Dundas, Bart and Dame E. Dundas. Included in the compensation received by the Hopetoun Estate Company is that for A. Hope, Bart. Included in the compensation received by the Coltness Iron Company is that for the Coltness Estate.

In view of these figures and the contemporary agitation for the public ownership of land, the royalties were a natural target for redistributing revenue. But the concentration of their ownership surely had other effects than consolidating the unequal distribution of income and wealth. Here the effects can be seen to be contradictory. For the ownership of large tracts of land meant a lessening of the problems associated with multiple leasing and spread of mines across many properties. On the other hand this potentially placed the landowner in a powerful position to extract a monopoly-type rent, as again can be illustrated by Scotland. For the relationship between the landowners and mine owners altered in Scotland during the eighteenth and nineteenth centuries. During the eighteenth century and up to the mid-nineteenth century, Scotland is notable for the degree of involvement of the great landowners in the running of the mines on their estates. Examples include the Earls of Wemyss in Fife, the Marquises of Lothian on Newbattle Abbey, the Earls of Rothes on the Leslie Estate, the Dukes of Buccleuch and Hamilton, and Erskines of Mar, and the Alloa Company. These landed families were prominent in the management and control of mining on their own estates in the Eastern Lowlands, in Fife, and in the Lothians especially, whereas in western Scotland 'tacksmen' or leaseholders were more common, especially in Lanarkshire and Renfrewshire. Smout (1964) believes that this difference stemmed partly from the nearness of Glasgow from which mercantile wealth could challenge the affluence of the land, as well as from the more modest resources of the lairds in Lanarkshire and Renfrewshire. The smaller gentry in the eighteenth century had less room to manoeuvre in estate management than had the great landowners. (See Thompson 1966 and Ward 1971.) For instance, the transport shortcomings in Lanarkshire were beyond the scope of investment for the small lairds in that county. Smout stresses that only large landowners could have afforded long-term investment without a view to immediate returns in land reclamation or road construction.

The growing scale of operations discouraged Scottish landowners from continuing to operate the mines directly from the middle of the nineteenth century.[15] Nevertheless, the greater concentration of ownership of the royalties relative to the rest of Britain explains the more active involvement of the owners initially and a greater persistence of involvement. The greater concentration of ownership is reflected in that the average number of leases per mine were 1.9, 3.0, 3.0 and 1.6 for the four Scottish districts in 1925, compared with 5 for the country as a whole. The Royal Commission on Mineral Royalties also reported that the power of the landlords was felt to be more obstructive and that stringent conditions governed the leases, which were in general shorter and with more limited powers of surrender and assignment than elsewhere.

These conditions concerning the ownership of royalties have to be considered in conjunction with competitive pressures. The geological conditions in Scotland were of the poorest and the industry survived only because of its relatively high level of mechanization. It was particularly dependent on export markets which were very depressed in the interwar years. This had the effect of maintaining the level of royalties, given the power of the landlords, but the mines still tended to be squeezed within the confines of a single estate. The result is that the process of concentration in the district did not match the progress in mechanization, with mine size growing relatively slowly compared to the British average, often from small pits in the first place (see table 4.5).

Table 4.5 Mechanization, amalgamation and royalties

	Output 000 tons		No. of mines		Output/ mine 000 tons		No. of mchnd mines		Propn. of coal cut mchnly		Royalty/ ton pence	
	1922	1938	1922	1938	1922	1938	1922	1938	1922	1938	1925	1938
District 1	9,284	8,288	68	51	121.8	162.5	49	44	43.6	89.4	7.14	
District 2	4,295	4,442	41	33	104.8	134.6	21	25	25.2	62.2	6.96	
District 3	18,583	13,057	313	284	59.4	50.0	165	132	23.9	47.1	6.63	6.3
District 4	4.286	4,493	91	58	47.1	77.5	21	37	18.4	69.4	5.57	
Britain	227,015	249,607	299	2025	78.0	123.3	785	927	15.3	59.4	5.69	

Source: Annual Reports of HM Chief Inspector of Mines

From private obstacle to public burden?

The problems of the royalty system then were both those of powerful landowners confronting different mine owners within their property (in pure form, a single landowner developing a number of competing mines) and of mine owners seeking to negotiate for coal across a number of competing landlords (in pure form, in the absence of other mine owners). Nor were these problems subject to a solution once and for all, because of the varying location of the coal to be extracted and the expansion in output and geographical spread of the mines. Indeed, the problems were to become more complex, a simple index of which is the number of leases to be negotiated by a single mine which increased between 1925 and 1938 from the five reported by Samuel to an average of between ten and fifteen to remove known coal reserves even on the basis of the existing fragmented and small scale layout of British mines.[16]

The most obvious difficulty lay in the leaving of coal barriers. The Royal Commission on Coal Supplies (1903–5) reckoned that as much as 9,500 million tonnes had been left underground for the support of the surface and in barriers, thirty times more than the maximum annual output ever achieved in 1913. Of this the Ministry of Reconstruction (1916/8) estimated that just less than half had been left in barriers alone. This is not just coal wasted pure and simple, for the barriers did perform a technical function, in preventing flooding, spread of fire, and, intertwined with this, the marking of boundaries between leases. But the latter was only necessary in order to secure each mine independently from adjacent mines against the risks mentioned. It is as if each household were to build its own garden fences in case a neighbour did not oblige, thereby duplicating the costs of fencing. As the Ministry of Reconstruction concluded:

> The total area of minerals left in such barriers is very large, and we are satisfied that a comprehensive survey of existing barriers would show a considerable portion could be worked with safety.
>
> (1918: 34)

Further, the Report pointed to the benefits and necessity of co-operation between those involved because of the presence of externalities:

> The problem of the working of barriers differs from that of the working of minerals left for the support of buildings or surface structures . . . *viz.*, that in the working of a barrier the damage which may result from its removal cannot be entirely gauged from the working of the particular barrier in question, but is contingent upon whether other barriers are left or worked – it may be be over a considerable area of the mineral field . . . . Mutual private arrangements have been carried out in some mineral districts in the way of taking over

pumping stations and concentration of the pumping of water. This has had the effect not only of economising in the cost of dealing with the water, but also of allowing a number of mineral barriers to be safely dispensed with, as, for example, in South Yorkshire.

(p. 23)

During the interwar period, problems such as these led each of the large number of official bodies and reports considering the industry to recommend unification of the royalties into a single ownership agency, with public ownership generally being the preferred form. But the nature of the argument had also changed. Much greater emphasis was placed upon the role of the unification of the royalties as a means of bringing about the reorganization of the industry. Thus, in its report for 1933, the Coal Mines Reorganization Commission observed that:

The Royal Commission (Samuel of 1926) recommended that coal royalties should be nationalised, and in 1930 the Government said that they intended without delay to ask parliament to pass such a measure. That the State should have the powers of a lessor over the whole industry seems to have been regarded by both as an integral part of the plan of stimulating reorganisation from the outside. The Royal Commission evidently intended that the authority which administered the State's rights of mineral ownership should be the principal agent in bringing about amalgamations. But royalties have not been nationalised. The vast powers that would be possessed by a universal mineral lessor cannot be called in aid.

(CMRC 1933: 4)

In the conclusion, the Report refers not only to the benefits for reorganization bestowed by the unified ownership of the royalties, but also points to the positive powers of determining who, and to some extent, how coal should be produced:

'What', we are asked, 'is the use of doing all these troublesome things to eliminate waste, secure planned development, and get rid of superfluous units, so long as so much still depends on the accident of mineral ownership, and our plans may always be stultified by the opening or re-opening of mines without regard to corporate efficiency or national need?' We can only answer this question by saying that Parliament will presumably remove sooner or later – whether by nationalisation or by some less sweeping reform – an impediment that so seriously obstructs the fulfilment of its policy.

(pp. 12–13)

From private obstacle to public burden?

No doubt the peek-a-boo behaviour of the smaller mines was particularly galling in the context of the fragmented structure of the industry and in face of a severely depressed market for coal. Sir Ernest Gowers, Chairman of the Commission, was to make similar remarks about royalties in a Lecture to Businessmen in Cardiff. He insisted upon some scheme of unification of the royalties, whether under public ownership or not. This was part of a campaign to win over intransigent mine owners to the schemes for reorganization proposed by the CMRC. In this, Gowers proved a very unpopular figure and the CMRC was ultimately defeated in its efforts.[17]

In the meantime, he had been busy behind the scenes preparing for the unification of the royalties. Working through a high-level interdepartmental committee in 1934 and 1935 considering how and whether to nationalize the royalties, he drafted and gained ministerial approval for a report recommending the principle of unification of the royalties, their administration by the CMRC, and including a proposal on how to finance compensation (Public Record Office, Coal 12/14).

Essentially, the CMRC's view of the royalties was that they could serve as a weapon for bringing about the reorganization of the industry, as a means of solving the problems of small-scale, unmechanized mines. Much less emphasis was placed upon the idea that they were part of the problem. At the same time the emphasis had changed, as previously remarked, to a different set of problems, from difficulties around the fringes of mines and properties to the structure of the industry as a whole, and to a concern for practicalities rather than with the revenue and distributional implications of private royalty ownership. Indeed, the CMRC's ethic, whether in pushing through reorganization or, more implicitly, in unifying the royalties, was that through the provision of appropriate compensation, no one should be made worse off.

In this light, the mineral owners and the royalty system appear more as a passive support to the industry's problems, even if an instrument of reorganization in the making. This is to underestimate the role of the royalty system at an economic if not at a practical level and, although this distinction has been here, to force a separation between the two. First, observe that royalty owners did make economic calculations – whether to mine coal or not themselves, this reflecting a choice over the allocation of finance between competing demands whether for investment on other land-intensive activities or in conspicuous consumption. They also negotiated leases, often incorporating complex conditions to ensure early and full reward for the coal extracted and even making claims upon related industrial endeavours. Here is a notable example for a 999 year lease drawn from evidence to the Samuel Commission:

> the rent stipulated was so much in case only one blast furnace was erected, a higher rent if two, a still higher rent if three and so much more for every additional furnace, it being understood that each furnace should not smelt more than a certain quantity of iron ore in each year but if more than that quantity is made in any one year the rent is to rise or be augmented in proportion to the quantity made compared with the quantity before specified.
>
> (Volume II: 768)

Second, the mineral owners were not passive but were organized as an interest group. In 1911, the Mineral Owners Association of Great Britain, MOAGB, was formed by Earl Fitzwilliam and others as a 'non-political alliance for the protection and benefit of persons receiving mineral rents'.[18] Up to 1919, it was concerned primarily with issues of dispute with the Inland Revenue. However, after the prospect of royalty nationalization following the unanimous recommendation of the Sankey Commission, the Mineral Owners' Joint Committee was set up in 1920. It comprised a variety of previously existing organizations apart from the MOAGB – the South Wales Royalty Owners Committee, the Central Land Owners Association, and the Mineral Owners' Committee of the Scottish Land and Property Federation. The Joint Committee ultimately gave up the fight against royalty nationalization only in 1936, once it was convinced of the unyielding governmental commitment to it. Making the best of a bad job, they then rallied around fighting for their terms of compensation, pressing for a global figure of £150 million. When £66.45 million was finally announced, reportedly, '. . .the mineral owners, though the award meant ruin to some and hardship to all, seemed helpless'! The size and the concentration of the sums involved suggest a luxurious ruin!

Third, it has been suggested by some that the royalties could not have been that important since they were such a small proportion of working expenses. Thus for Church (see also Dintenfass (1985: 369 and 1988)):

> when royalty estimates are compared with estimates of profits, they show that while royalties per ton actually exceeded profit per ton until the mid-1890s, thereafter until 1914 profit per ton ran at a level roughly twice the average of royalties; moreover, throughout this period royalties as a proportion of working costs averaged barely 5 per cent. Royalties, therefore, were less of a burden during the twenty years before 1914 than hitherto; moreover, the higher profit levels of this period suggest that there was no shortage of capital attributable to a drain into royalties.
>
> (1986: 775–6)

This raises a number of separate issues. One is of timing. For reasons given previously, for smaller mines in a less developed industry of lesser age in the environment of a system of large landed property, there is the potential for landed property to stimulate flow of capital into the sector. That this should lead to problems around the borders of the industry over the period before 1914 is not contradicted by the aggregate evidence on royalties and the comparison with costs and with profits. Indeed, it follows with expanding output that royalties would also have been increasing if not, as it happens, as fast as profits. Also as argued at greater length in chapter 5, the significance of royalties cannot be measured by their absolute or relative size in revenue terms. The comparison with profits is instructive for it might also be argued that these were also not important as they were such a small proportion of costs (although not costs as such themselves) – but they are the motive of the whole enterprise as far as capital is concerned.

Equally, it would have to be argued that capital costs were not so important as they were small when compared to the overall wages bill. It is worth observing that the price of nationalizing the royalties was as much as one-half the amount paid for taking the industry's mining assets as a whole into public ownership once, for the latter is excluded compensation for ancillary investments such as brickworks, railway trucks, miners' housing, etc. For in 1946, the total cost of nationalizing coal and royalties and other assets was £394.4 million. Of this, the cost of the royalties now made up £80.9 million and the coal industry assets £164.7 million (Ashworth 1986: 28), although the royalties included in principle all workable and not only currently worked coal. By this comparison, Church's argument could be used to turn his conclusion upside-down, making the royalty system as (half as) important as the universally recognized inefficiencies in the structure of investment!

Fourth, most important, however, is the argument that the major impeding role of the royalty system would occur in the interwar period, once the industry was entering the age of mechanization and amalgamation. This could only serve to intensify the technical, financial, and legal problems that undoubtedly existed earlier around the royalty system. But this cannot be understood purely as a worsening of practical problems alone. For this is not a once and for all shift for which comparative statics best describes the new equilibrium with enhanced technology in place of the old. Rather the process of change is one in which the individual capitals expand in size through accumulation. For mining, an inescapable condition for them to do so is through access to greater stretches of land, whether in the expansion of existing mines or in the setting up of new ones.

Consequently, landowners will attempt to enhance their royalties by appropriating a portion of the surplus profitability that would otherwise accrue fully to the mine owner. The result is that the motive to accumulate is dulled through diminished rewards but that royalties do tend to be higher where capital investments do take place. This argument is unaffected in principle, should the industry be in decline, for then mine owners will seek to minimize their reductions in profitability and to preserve their continuing operations (not write off capital expenditure, for example) so that in absolute terms overall, this is consistent with declining royalties and with landowners making concessions over them. Indeed, this is part and parcel of a system of royalty payments sliding with the price of coal or profitability.

Elsewhere in this book, empirical evidence is brought forward from the interwar coal industry that is consistent with this sort of analysis. It is also set unfavourably against neoclassical theory and the residual hypothesis of entrepreneurial failure. Unfortunately, however, a more positive empirical analysis is beyond the scope of the research covered by this book. As is clear from the variety of issues covered in this chapter, the exact manner in which the role of landed property emerges is extremely complex, reflecting technical, legal, geological, and financial conditions quite apart from the juxtaposition of the ownership of coal seams and the progress of individual mines. This makes detailed individual studies a prerequisite of further progress. Hopefully, other scholars will be encouraged to do this, taking a greater interest in the role of landed property than hitherto and, in this respect at least, matching the concern of those charged with reorganizing the industry at the time.

Concluding remarks

This and the last chapter have in part been constructed as a dialogue between two different approaches to the royalty problem. One has concerned itself with theoretical problems in which royalties are seen as a revenue, either smoothly and harmoniously bringing about the efficient allocation of resources or acting as a distributional drain on the resources available for wages, investment, or price reductions. The other focuses on practical problems in (re)organizing the industry, in bringing about amalgamations and minimizing the costs of pumping water and of coal lost in barriers.

Ultimately, it has been concluded that this distinction between the economic and the practical is not tenable. The appropriation of revenue by landlords as a condition of access to the land has an impact on how the mining of coal is organized, just as this in turn affects the level and direction of the royalties paid. The exact relationship between the two is complex and varies over time, incorporating legal, financial, as well as technical considerations – indeed potentially any factor that is independently influential on the industry's progress. For, in so far as these differentially affect the productivity of (potential) mines, the subsequent profits are the target of landlords who may encourage and support such developments in so far as they are guaranteed a return for doing so. Otherwise, landed property may act as an obstacle in that the profits that would otherwise accrue to dynamic entrepreneurs are denied to them.

Clearly, the view offered here is that these two possibilities followed each other historically in the order presented, with the turn of the nineteenth century acting as a watershed between them. Moreover, the problems of the second phase, which would have been confronted in any case by the spatial expansion of mining that inevitably occurs as initial reserves are exhausted, were accelerated by the pressures to mechanize and rationalize the industry. Hence the almost unprecedented pressure to violate the principles of private property, leading to the nationalization of the royalties in 1938, some hundred years or more after it had been necessary elsewhere in Europe, but under a Conservative Government and leading the major nationalizations of the first majority Labour Government by the interval of the Second World War.

How is all of this to be understood theoretically? Clearly not within orthodox theory with its reliance upon static equilibrium models, nor within the radical alternative which merely seeks static redistribution. Marx's theory of agricultural rent does offer an alternative starting point.[19] First, it emphasizes the necessity of examining the specific historical circumstances in which capital confronts landed property. Unlike neoclassical, Ricardian, or radical distribution theory, rent does not simply fall out from technical conditions of production governing the application of capital to the land (with the radical theory also appending a monopoly price for the use of land). Rather, Marx's concern is with the specific economic and other conditions which govern capital's access to the land.

Second, in the context of differential rent (of the second type), Marx specifically focuses on the differential profitability that arises out of the differential application of capital intensity and the barriers to accumulation that this may yield if such excess profitability is appropriated by the landowner rather than the capitalist.

Finally, in the context of absolute rent, Marx's concern is with the application of new additions of capital to the sector. He draws the conclusion that this must draw a rent even on the worst land in use and that together the processes of investing on existing land and on new land can lead to an underdevelopment of the industry in terms of the degree of mechanization.

The parallels between this theory and the experience of the coal industry are not accidental, since the research reported here was itself the outcome of a dialogue between Marx's theory and the development of the British coal industry in the attempt to understand and interpret each of them. But possibly the analysis is of wider relevance. For recently, interest in Marx's theory of rent has been revived in the context of urban crisis. In this Marx's analysis has often been treated as if it simply concerned distribution of surplus value between landlords and tenants (with other distributional categories such as capitalists and workers occasionally added) and has otherwise been rejected as being too simplistic and empirically invalid.

This has led, in particular, to naive and backward economic and political analysis in the UK in recent years, especially concerning the housing question. This issue has increasingly been reduced to one of the forms of housing tenure (most notably owner occupation versus council tenant) rather than to a construction of a fuller analysis of the structure of housing provision (building firms, access to building land, work organization, terms of finance, etc.).[20] Accordingly, policy becomes oriented around whether one form of tenure receives a greater subsidy than another, i.e. as a distributional matter alone.

But to return to the royalties. They were encompassed within a new form of tenure following their nationalization in 1938, although this has been overshadowed in the historical memory by the post-war nationalization of the industry. The royalty owners were gone but their compensation still needed to be paid. Without any rationale, the cost of compensation was treated as a capital liability of the newly formed National Coal Board and placed an interest burden on the public corporation that had to be paid to the Government. The royalties as an asset were finally written off in 1973 at a value of £52.4 million. The original compensation had been fixed at £66.5 million and this had swelled to £80.9 million by 1946. Consequently, the NCB had paid off £24.4 million together with interest over a twenty-seven year period.

This was a heavy burden on an industry that was kept short of investment finance. The result is a remarkable historical irony. Towards the end of the nineteenth century, the miners began to press for the public ownership of the mines and the royalties, for the latter in the belief that this would relieve the industry of a parasitic cost. Instead, after defeat in the General Strike, neither nationalization was achieved and the revenue of the industry was sustained in the 1930s by a state-organized cartel, thereby funding both the costs of the royalties directly and the indirect costs of the royalty system in its effects on the industry. After nationalization of both mines and coal, compensation for the latter was borne as a cost by the NCB. What the miners had hoped to abolish was paradoxically created by public ownership!

Part three
Cliometrics and coal

Chapter five

Returning to factor returns: the late nineteenth century coal industry

In an early article within the New Economic History of Cliometrics, McCloskey (1971) has attempted to demonstrate that differences in productivity between the coal industries of the United States and Britain can be explained by differences in natural or geological conditions.[1] In so doing, he exonerates British entrepreneurs from failing relative to their American counterparts. The purpose of this chapter is not to put the British entrepreneur back into the dock to stand trial as once again accused of failure. Indeed, most of the analysis presented here would remain valid and relevant even if McCloskey's calculations had proved most unfavourable to the British entrepreneur. Rather, the purpose is to expose the general limitations of the method employed by McCloskey as well as some of the more specific criticisms that arise because of its application to the British coal industry prior to the First World War.

The critique begins in the first section by presenting some of the problems of estimating entrepreneurial performance through the calculation of total factor productivity. In the second section, the discussion moves towards some of the issues that arise specifically out of the application of the method to landed property. In the course of doing so, it is seen how landed property may have played more than a passive role in the development of the industry, rather than its being confined to a source of revenue in correspondence to fertility differences (and thereby guaranteeing equalized rates of profit and efficient allocation of resources across mines, as suggested by neoclassical theory).

Thus, whilst the contribution of this chapter is primarily negative, the beginnings of an alternative will also be presented, and this will perhaps serve as a guide to those undertaking more detailed studies of the coal industry or other sectoral studies in which landed property is closely involved. Indeed, the analysis here conforms with the results of the previous two chapters where it has been argued that neoclassical economics has become incapable of constructing a specific theory of the role of landed property, just as the theory of rent has been more or less completely neglected.[2] It should come as no surprise that the empirical

71

counterpart to the neoclassical theory of landed property should prove equally unpalatable.

General problems

The method used by McCloskey is as follows: suppose that the United States and Britain have the same production function for coal. Is it possible to explain different levels of labour productivity by the different per capita inputs used? If the answer is in the affirmative, then McCloskey deduces that there is no entrepreneurial failure, and the masters and men of British coalmining will have been unjustly accused even if only achieving lower levels of productivity.

Even at this general level there are still problems. It is, of course, possible that both US and UK coal industries failed. The comparison would not reveal this. Equally, if the two countries did not share the same production function, then to draw conclusions on the basis that they do is extremely precarious. McCloskey is aware of this but is content to presume that imitation of technology – through whatever route – is sufficient to render the assumption a reasonable basis on which to proceed.

Here, a particularly narrow view of what constitutes the production function becomes clear. In principle, any factor that influences production could be included as an input affecting output (and the more that are included the lower is the unexplained residual that is 'explained' by technical progress). In practice these factors are to be limited precisely to those traditionally conceived of as inputs – labour, capital, and land in the narrowest of technical senses. The particular relations between the classes that own these factor inputs are rendered irrelevant except in so far as they come to the market place to sell what they own. Yet, it is precisely such relations between classes that are the least mobile as between countries, whatever the mobility generated through the transfer of technology. This is the light in which should be placed Lazonick's criticisms of the application of the factor productivity method to the cotton industry. His emphasis is on differing industrial relations as between the UK and the US. These are no doubt of equal importance to the coal industries, but the focus here will be on the role of landed property.[3]

Before proceeding to this, it is instructive to examine other assumptions that are necessary to warrant the adoption of the factor productivity method.[4] The first of these is that there are constant returns to scale. These are necessary to guarantee that factor shares exactly exhaust output so that, in perfect competition, marginal products can be measured. For some, constant returns to scale must necessarily prevail if all inputs are exhaustively included. But this is unconvincing since it is always possible in principle to duplicate what has previously been achieved and it may be possible to do better. Examples are easy to construct – a boiler increases its volume and capacity as the cube of its dimensions whilst the demand for materials to construct it only squares the dimensions. More to the point, a simple shaft to a mine could serve a number of seams, workings and roadways. To the extent that there are large capital outlays necessary to get to the coal even prior to working it, economies of scale appear to be endemic.[5]

Significantly, McCloskey gets round this problem by suggesting with minimal evidence that capital per worker may have been about the same in the two countries. In the UK, the capital:labour ratio may have been higher because of the difficulty of working coal, given its greater depth for example, but also have been compensatingly lower because of a lower relative wage with a reduced incentive to substitute capital for labour. As it were, the UK substitutes capital for land and the US capital for labour.

With the now presumed equality of capital:labour ratios and constant returns to scale, this allows the role of capital to be set aside altogether. Of some importance, however, is the different types of mechanization, from haulage to coal cutting, as well as the infrastructure of the mines. It is surely inadequate to lump these together and presume they are being equally efficiently used when the weight of reported evidence, as noted by McCloskey, is to the contrary.

A second assumption required by McCloskey is that of perfect competition. This guarantees that marginal products are defined and equal to factor rewards. But even a casual inspection of the pre-war coal industry suggests that economic conditions are otherwise. As Taylor (1968) observes there are pronounced cyclical movements in tune with the rest of the economy but with greater amplitude. The industry's cycle appears to have been associated with a distributional conflict between capital and labour, with wages (and absenteeism) favoured as the labour market tightens. Certainly, the picture is not one in which overall demand considerations can be ignored, nor one in which class conflict, as a source of distribution, can be set aside.[6]

A third necessary assumption is that factor rewards measure corresponding contributions to output and this is a heroic assumption even if constant returns and perfect competition are permissible. A proof of the correspondence depends upon there being a single good in the economy and the fact that McCloskey also considers steel in his article is an eloquent illustration of the fallacy of such an assumption. For those familiar with the Cambridge critique of capital theory, the errors in constructing production functions by appeal to factor rewards will be well-known. It can be shown that beyond the hypothetical one-good world, the measurement of technological performance by factor rewards confuses price (and distributional) changes with production changes. If wages are higher in the US, as compared to the UK, this has the effect of changing techniques of production (as is supposed to be underlying production function theory) and all associated prices which are then used to measure factor rewards.

The Cambridge critique is associated with the idea that changes in wages are the result of distributional conflict between capital and labour. If so, measuring contributions to production by factor rewards will measure the two things – production change and distributional change – as if they were the one, that is the former. Consequently, a change in distribution alone will appear as a change in production.[7] Crucially, however, the argument against the measurement of production functions in this way does not depend upon acceptance of the Cambridge view of the prior determination of distribution. Exactly the same considerations apply if the role played by demand is recognized. On the assumption that demand for coal in the UK is different from that in the US, then these differing demands affect factor rewards as much as the conditions of supply. To measure factor rewards as if they were exclusively determined by the conditions of supply is to commit the simplest if, in this context, commonest of errors.

Moreover, general equilibrium theory informs us that the simplest intuition in these matters can be fallacious, so perverse can price movements be in all but the most restrictive conditions.[8] For example, an increase of wages in the US as compared to the UK may very well make more capital-intensive industry efficient in the UK once all demand considerations are taken into account. The conclusions to be drawn are simple and devastating. The production function methodology is logically unacceptable. It measures changes in supply and demand as if they were exclusively changes in supply. By doing so it will calculate effects not only of the wrong magnitude but even of the wrong sign. The method gives no accurate indication of what it purports to be measuring – entrepreneurial performance via factor rewards.[9]

Even putting these objections aside, there is a peculiarity in the total factor productivity method. It is in the conclusion that the contribution to production is measured by factor rewards in the presence of perfect competition, constant returns, and a single good. If this were generally valid, it would in principle render the economic historian's task an extremely simple one. The contribution of such and such to the economy equals the revenue that such and such received. We might find this acceptable for what are primarily economic agents, but it is clearly facile for non-economic agencies, such as the government. Whatever the contribution of government to the economy few would reckon it in terms of the revenue it received or spent. Indeed, the neoclassical method that supports the microeconomic measurement of production functions is often associated with a monetarism in which the state's contribution is in inverse proportion to its expenditure!

The reason for this paradox is that the government is not primarily a market institution even if it does have profound effects on the workings of the market. But, by the same token, it is necessary to recognize the non-market role of what might, nonetheless, be considered primarily economic agents. Indeed, it is not clear that the contribution of economic agents through the market is entirely separate from non-market factors. The market of perfect competition is a complete fiction. In reality, the market performs its role in definite social, political, and legal circumstances and these have an effect that has to be determined and over which economic agents attempt to exert an influence. It also makes the job of the economic historian more interesting than that of an accountant of revenues.

Particular problems of landed property

These general remarks are intended as a preface to a more specific consideration of the role of landed property in the US and UK coal industries, with focus primarily on the UK. For the production function approach, the matter is simple enough. It is reduced to a question of geological conditions. The way in which the land is owned and the problems of ownership do not figure. Set this aside for the moment. McCloskey attempts to explain table 5.1 by differing mining conditions.

Table 5.1 Yearly output per man employed, UK in 1907 and US in 1909

	Output (millions of tons)	Employees (millions)	Output/man per annum
UK	267	0.812	325
US	408	0.667	613

Source: McCloskey 1971: 301–2

He begins by estimating the supposed effect of available reserves, reckoning these to be five times greater per worker in the US than in the UK. As the distributional share of land is approximately 8 per cent, this would explain a productivity difference per worker of 8 per cent × 80 per cent or 6.4 per cent, since the percentage difference in coal per worker is 80 per cent.

This is a most peculiar procedure. First, as is well known for most extraction industries, proven reserves are an endogenous variable which tends to keep well ahead of minerals being worked. It is as such an inappropriate measure of the land factor input. Even so, the level of reserves per worker for the UK is in the region of 100,000 metric tons. Increasing this to 500,000 as for the US worker can hardly be anticipated to increase productivity. Placing the land:worker ratio at a level for UK:US of 0.205:1.000, as is warranted by constant returns and as McCloskey does, merely tends to conceal the massive reserves available to both countries irrespective of those yet to be discovered and/or counted.

These considerations are to be kept quite separate from the quality of the coal land available, which would be increased with proven reserves in general. McCloskey goes on to consider the effect of seam depth and thickness on labour productivity. He finds that the deeper, thinner seams of the UK are sufficiently disadvantageous to explain the remaining productivity differences between the US and the UK. Admittedly, the exercise is crude but, quite apart from the calculations, the question is what does it demonstrate? The answer is that the UK mine owners worked deeper and thinner seams with an estimated effect on productivity. Yet, McCloskey himself observes that the seams worked are a matter of choice, thicker seams compensating for greater depth. This skirts the issue of whether mine owners worked the best seams that were available and did so in the most efficient manner. Since the seams worked are not purely a geological property and are subject to choice, the performance of the industry must be judged on the basis of those choices. On this McCloskey is silent, although there is perhaps the presumption that the market will have worked perfectly to make those choices efficient.

There is much evidence that the system of landed property in the UK made these choices inefficient. To confirm this absolutely would involve a detailed examination of the industry and it is far from clear that the information is available to do this. More indirect evidence can, however, be used. It does suffice perhaps to suggest that the working hypothesis should be one of an inefficient use of land in the UK that worsened towards the end of the nineteenth century.

In the US, the mine owners in general owned large tracts of land over which they could conduct their business without impediment. Given the superior geological conditions, it is understandable that mines were shallower and often smaller than in the UK. Here, although the same law applied to landed property, there was a separation of ownership between the minerals and the mines. Moreover the ownership of the coal royalties was extremely concentrated, reflecting the concentration of ownership of land in general. For example, in the four Scottish districts, the ten largest royalty owners accounted for 75 per cent, 70 per cent, 43 per cent, and 70 per cent of all potential coal upon nationalization of the royalties in 1938. For the country as a whole in 1918, a mere 100 royalty owners received almost 50 per cent of the royalty income and there were at most three or four thousand royalty owners altogether.[10] This evidence alone is sufficient to cast doubt on the idea of perfect competition yielding efficient choice of seams to be worked particularly when it is borne in mind the length of time for which a mine is worked and that it is heavily tied *in situ*. Landowners had tenants' capital strongly tied to their property in the form of the fixed capital irretrievably sunk in the infrastructural fabric of their mines.

There were also considerable practical problems confronting the presumed perfectly working market. The Acquisition and Valuation of Land Committee, under the chairmanship of Sir Leslie Scott, which ultimately gave rise to the Mines (Working Facilities and Support) Act of 1923, listed fourteen problems with the royalty system, the best known of which being that of barriers, as discussed in chapter 4, the leaving of coal unextracted to mark boundaries between mines and in conformity with surface leases (all of which might still be owned by a single landlord). More coal would be liable to be left in those mines whose boundaries did not lie within a stretch of land owned by a single individual, for which the removal of boundaries is less easily negotiated. Thus, the problem of lost barriers would tend to increase with the expansion of the industry through mines encroaching across land held by different landlords.

The Scott Report's thirteen other such problems were predominantly concerned with difficulties associated with the division between property rights. As it had taken ten years to negotiate one of these problems (subsidence) between interested parties, a certain impatience on the part of those involved may be appreciated even if, in principle, other difficulties could have been resolved through some long-term compromise. The negotiations over subsidence had taken so long and were so delicately balanced that it was simply insisted that no amendments could be made to the enabling legislation as it passed through Parliament. Details of the history of this legislation are to be found in the Second and Final Reports of the Royal Commission on Mining Subsidence (1926–7). In Parliament, Scott suggested that:

> the agreement contained in Clause 15 of the Bill was really almost like a treaty made between the interests affected after prodigious discussions that had lasted from 1919 until last autumn, when agreement was finally reached as between the two parties.

The Subsidence Commission dates the problem from 1913 when a court decision (Howley Park) proved the existing law inadequate. But the problem again is one of property rights in face of a spatially expanding industry. British mines were becoming deeper and, for this to be cost effective, more widely spread. As a result, subsidence became a more serious problem and was no longer physically confined to the area immediately above the extracted seams as the mine's roof was allowed to collapse. Previously, the law concerning damage had been constructed on the basis of belief in such vertically confined damage. The new conditions laid the mine owners open to considerably larger claims for compensation, especially around railway tracks.

Yet another problem concerned the 'externalities' governing the draining of water for which efficient organization of dams and pumping were impeded by the need for drainage systems to conform to the dictates of property rights. A further problem, not mentioned by the Scott Report, but of importance because of its relation to the control of labour, concerned the premium for large coal. It is important to recognize that coal is not a homogeneous product as it yields a vast variety of different characteristics, including size. At this time, and into the period of nationalization, there was a price advantage in large coal. In relation to the royalty owners, the mine owners had an incentive to extract large coal and even leave the small to the extent that royalties were paid by the ton or some form of equivalent. By the same token, to the extent that the miners were on piece rates, they would fill tubs with the more easily available small coal where allowed and, if fined for doing so, would tend to neglect small coal. As piece rates were prevalent with fines in one form or another, the result was the wastage of small coal.

In this light, it can be seen that, until the nationalization of the royalties in 1938, the UK industry was belabouring under an inappropriate system of land law. The Act of 1923 was supposed to ameliorate the various problems without going to the extreme step of public ownership. But were these problems simply ones of practicality, what for neoclassical economists would involve high information and transaction costs for which public ownership might even be an efficient solution? This indeed is the sort of approach offered by McCloskey with his suggestion that the land in the UK was so valuable as to warrant separate ownership of surface and minerals despite the transaction costs involved:

> the pattern of coal land tenure is a good example of the effect of economic conditions on legal arrangements. In the United Kingdom, apparently, land was expensive enough to overcome the high transaction costs of selling mineral rights and surface rights separately and to warrant more specialization between ownership of the rights and exploitation of the rights.
>
> (1971: 301-2)

Here it is simply assumed that the patterns of specialization were in an efficient correspondence with economic conditions. It is true that the economic conditions initially dictated the legal condition of private ownership of the coal royalties, although this resolution of Elizabethan conflict between royalty and landowners some four hundred years previously is not presumably what McCloskey has in mind. Subsequently, it is the legal conditions that affected the economic conditions, and adversely, rather than vice versa. In short, the orthodoxy simply assumes away the possible effects of landed property and even takes as empirically supportive the separate ownership of coal royalties whilst, even in neoclassical terms, the transaction costs appear to have become too high to warrant separate sale of mineral and surface rights.

In this light, it seems as if there were significant differences in the systems of landed property in the UK and the US quite apart from differences in geological conditions. These differences in turn appear to have had significant effects on the potential for organizing mining efficiently and to have laid the foundation for poor layout, haulage and transportation, certainly by the end of the interwar period as judged by the Reid Report (1945). It is, however, also plausible to suggest that the system of ownership of landed property was related to certain obstructive economic mechanisms, once analysis is moved outside the fiction of perfect competition.

Picture mine owners making use of greater capital as it becomes available through self financing and/or borrowing. In investing, they will seek increasing profitability and may be able to obtain it given certain economies of scale. On the other hand, landowners will be able to demand some of this profit as a condition of access to the coal. As a result, the incentive to large-scale investment will be diluted since mine owners have to share the associated increases in profitability. To a greater or lesser extent, competition will force a compromise between land- and mine owners according to geological conditions, for example. Landowners overall may not even gain large increases in royalties but the system operates to discourage an efficient organization and reorganization of the industry. Returning to our earlier theme, the effect of the factor land on production cannot be reduced simply to its share in revenue, as a reflection of differential costs.

This is illustrated by table 5.2 which crudely tests mining conditions against royalties. If average width of seam cut is taken as a (poor) proxy for natural advantage, then there is if anything, leaving Wales aside, an inverse relation between this and the level of royalties. On the other hand, there is no systematic relation between the level of royalties and the percentage of mechanization. Overall it seems as if royalties are at best random and at worst perverse!

However, suppose that individual royalties increase relative to others with mechanization, whilst they initially reflect natural conditions to a large extent, but that mechanization only occurs on the worst mines first as competitive pressures force co-operation between mine and royalty owners to allow survival. This would yield an inverse relation between mechanization and width of seam (as indeed there is) and a complex relation between royalties and width of seam according to the extent

Table 5.2 Coal royalties and fertility compared

District	1925 Royalties/ton in old pence			1924 Average width of seam cut Feet	% Coal cut by machine
	Lowest	Highest	Average		
Scotland	1.50	26.50	6.68	39.55	46.4
Northumberland	3.00	12.42	5.74	41.83	26.7
Durham	2.21	24.00	6.32	45.26	16.0
S. Wales and Monmouth	2.00	21.00	7.83	53.08	5.4
S. Yorks	1.35	14.00	4.38	57.07	10.5
W. Yorks	1.35	9.75	4.85	43.37	24.2
Notts and Derby	1.20	10.05	3.90	50.97	14.8
Leics, Worcs and Cannock Chase	0.50	15.18	3.72	62.33	15.0
Lancs, Cheshire, and N. Staffs	0.50	20.23	4.84	52.01	16.7

Source: (Samuel) (1926) *Report of the Royal Commission on the Coal Industry*, Cmnd 2600, II: 768.

that mechanization has proceeded. Two polar extremes illustrate this, Wales and Scotland. Both have high royalties but each is towards one or the other extreme of both seam width and level of mechanization.

The processes described in the previous paragraphs can be viewed in another way in response to the question: why did the mine owners not buy the land that they worked? This might be particularly expected as landowners were quite heavily involved in the early nineteenth century as mining entrepreneurs but had almost completely withdrawn by the end of the century.[11] This is usually explained by the scarcity of finance for other purposes possibly also related to use on the land, such as farming. Thus, in the absence of a perfectly competitive market for finance, in which unlimited quantities are available to all at the going rate of interest, royalty owners could benefit from the finance available to others as capital is accumulated in mining. As capital availability increases to mine owners, so does potential profitability upon its application to the land. The associated increases in rents simultaneously increase the price of land as the discounted present value of future rentals. So the same process that makes capital available to purchase land also tends to increase its price. It is considerations such as these that would appear to explain the great stability of patterns of landownership as opposed to notions of perfectly working markets for which the patterns of ownership are more or less irrelevant.

Concluding remarks

McCloskey made an early and important contribution to the New

Economic History. As he himself locates his analysis in contrast to an old economic history that might be compared to an after-dinner conversation made hazy by brandy and cigars and suffering from speculation and loose logic, it is worthwhile to conclude here by assessing McCloskey's contribution in the wider context of cliometrics as a whole. It sets itself the task of making its underlying theory systematic and rigorous whilst confronting theory with the available empirical evidence. This is undoubtedly to be welcomed but in much the same way as virtue is a good thing. It all depends on what is understood to be virtuous.

Here it has been reiterated that the technique of production function estimation, much associated with cliometrics, is simply invalid. This is a well-known proposition within economics but has proved a limited deterrent there and there is no reason to expect any greater caution within economic history. This difficulty with total factor productivity, the Cambridge critique, holds even where the assumptions of and for perfect competition are thought to be acceptable. Of course, such assumptions are not generally acceptable, especially in those circumstances that are the subject of economic history. Thus, in exchanging the new history for the old, the increased level of rigour is open to question and, in any case, depends for its rationale upon assumptions that often border on the absurd. Balancing angels on a pin is made no more palatable by removing the topic from the realm of after-dinner speculation to the mathematical techniques of the computer room.

In the same breath, it must be recognized that the new cuts away much that is important within the old. This is particularly true when the new is reduced to statistical estimation of hypotheses loosely drawn from equilibrium models. What is cut away is precisely the history, for models are ahistorical, treating different circumstances as though they were the same. The basis for doing so is usually the presence of the market in common (and this also tends to provide for the availability of statistics). But the market works in different ways according to the historical circumstances in which it operates. Hopefully, it has been shown how this is important for the case of landed property in the UK as opposed to the US and over the history of the coal industry in the UK, with large landed property first acting as a stimulus then as an obstacle to progress.

Of course, cliometrics has not been confined to the estimation of production functions. In his survey of the field, Crafts (1987) chooses only one issue to cover, investigation of the Habbakuk thesis, in which production functions appear to be prominent (as opposed to unemployment, heights, demographic transition, and the general equilibrium modelling of the Kuznets curve). This is probably less representative of the weight of work done than of the sources from which the more interesting studies and techniques emerge. However, as an economist who has shown an interest in economic history, rather than vice versa, it is easy to observe that the New Economic History represents a repeat of the developments that took place in economics after the Second World War. The subject was made mathematically rigorous and the techniques of statistical enquiry have been prodigiously developed in the field of econometrics, especially recently with the enhanced and cheapened availability of computing. Theoretical rigour (and this is not just mathematics) and empirical investigation (not just econometrics) are indeed virtues, but they must not be advanced at the expense of historical content. To reproduce the shortcomings of economics would surely be to render economic history a truly dismal science.

Chapter six

Returns to scale in the interwar coal industry

The purpose of this chapter is to establish that there were considerable economies of scale in coal-mining in the British interwar industry.[1] This was certainly believed to be so by many commentators at the time who consequently pressed for amalgamations. However, an influential assessment of this matter by Buxton (1970, 1972a, 1972b, 1978, and 1979) has questioned the significance of economies of scale. He argues that these were not important since large-scale mines did not reveal higher levels of productivity. Rather, the source of productivity increase was to be found in the mechanization of mines, particularly coal cutting.[2] Further, the evidence brought forward to support economies of scale was unconvincing and therefore the belief in the benefits of amalgamation relied more upon conviction than upon serious analysis.[3]

In the first section it is shown that there are errors in Buxton's assessment of the importance of economies of scale. This is done in two ways. First, his own suggestion is employed, namely estimating the relationship between output per man, percentage of coal cut mechanically, and average size of mine. However, his method is found to be erroneous and his data set is extremely limited. Secondly, the relationship between output and factor inputs is estimated more directly, treating coal-cutting machines, labour, and the number of mines as inputs.

In each of these cases, contra Buxton, it is found that economies of scale are important. In addition, the second method finds that excess capacity was also an important factor in containing productivity increase. However, to reject Buxton's conclusions concerning the existence of economies of scale is to re-open the debate over the development of the interwar coal industry. For Buxton excuses the limited extent of amalgamations in production on the strength of his finding that economies of scale were limited. In so far as this is rejected, then the question is posed of why amalgamations did not take place. This is taken up in the final section.

Estimating economies of scale

Buxton's arguments concerning the explanation of productivity increase in the interwar coal industry are drawn from two sorts of correlation. The first shows a positive and significant relation between productivity and the degree of mechanization as measured by the proportion of coal mechanically cut; the second shows no significant relation between productivity and size of mine. These correlations are made from a number of different sources, for example, cross-section between districts or time series within a district. The results are used to conclude that mechanization was important for the industry but that economies of scale were not. There is, however, a simple fallacy in this reasoning. If productivity were determined simultaneously by mechanization and size of mine, and if small mines were more mechanized than large, then Buxton's correlations would be correct but his conclusions would not. Size would appear not to affect productivity simply because large mines were prejudiced against by their low level of mechanization. Ironically, Buxton argues that Scottish mines proved his point. They were small-scale but enjoyed high productivity because of their early and high level of mechanization, an advantage that was gradually eroded over the interwar period. Rather than supporting his case, these observations tend to suggest that economies of scale have been unfairly assessed. To correct Buxton's errors in statistical inference is relatively simple. Productivity per man must be explained simultaneously in terms of mechanization and size of mine. Rather than taking two separate simple regressions, a single multiple regression should be estimated. The results of these regressions are presented in detail in appendix 6.1. For eighteen districts[4] Q/L was regressed upon Q_M/Q and Q/N (using time series for 1921 to 1938) where Q is output, L employment, Q_M mechanically cut output, and N number of mines, each variable being per district per year. A time series analysis by district was preferred to cross-section by years since it was felt that mining conditions within a district across a time period would be more uniform than across districts at a given point of time. The results were conclusive. In all but five districts, both mechanization and average mine size were significantly positive at the 95 per cent level in explaining the level of productivity. Of these five districts, three found mine size insignificant, one mechanization insignificant, and one found both insignificant.

Buxton's single regressions were also calculated. When taken alone, mechanization was found to be significant for productivity for all eighteen districts. Contrary to Buxton's findings, all but fourteen districts found average mine size to be significant in explaining productivity. Nevertheless Buxton's relative ranking of mechanization and mine size was found to be correct in all but three districts, even if his absolute rejection of the significance of mine size is not accepted. Further, it was found that in eleven of the eighteen districts, the inclusion of mine size with mechanization improved the goodness of fit in explaining productivity levels. In summary, despite the shortcomings of this analysis,[5] it suffices to demonstrate that economies of scale cannot be dismissed as unimportant simply because mechanization is found to be significant. It must also be borne in mind that statistical analysis of this type is biased by taking the 'living' as representative of the 'dead'. Suppose size were important and competition did eliminate small-scale producers *ceteris paribus*. Then no correlation would be observed between size and productivity since the condition of survival of small mines would be their dependence upon some other advantage of which geological and other factors are bountiful in coal-mining. One could as well take surviving buildings as representative of the sturdiness of medieval construction.

The analysis of the effects of mechanization and economies of scale given above is relatively crude, with the proportion of coal cut mechanically standing as an index for the use of capital, the average mine size by district as an index of scale, and labour productivity as an index of efficiency. It is more usual to employ a production function in which the level of outputs is explained by the inputs used together with technical progress. Ideal would be to investigate the difference these factors make mine by mine as geological conditions remain unchanged. This is quite impossible because of lack of data. Accordingly, as above, each district is treated as if it were a single mining enterprise. It employs certain quantities of capital and labour spread over a number of mines. It is possible to investigate whether there were economies of scale by asking whether output would have been greater if the number of mines had been smaller after taking due account of the amount of capital and labour employed.

The method for doing this is to estimate a production function for each district of the form $Q = A K^\alpha L^\beta N^\gamma$. If γ is negative then output would have been greater for fewer mines. It could be argued that this might be the result of movement on to worse mining conditions, by analogy with the Ricardian extensive margin, as output is expanded. This, however, can also be investigated by asking whether output increases more than in proportion for a uniform increase in capital and labour with the number of mines remaining constant. This corresponds to the Ricardian intensive margin and supports the case for economies of scale if and only if $\alpha + \beta$ is greater than one.[6]

It has been traditional to estimate production functions by measuring marginal products from factor rewards. It has been shown, however, in the context of the capital theory debate, that this method is erroneous[7] and so it was not used. In any case, the method requires constant returns to scale and full employment of resources for marginal products to fulfil their assigned role, conditions which are both inappropriate for the interwar coal industry, and an investigation of the presence of economies of scale. Instead the production function was estimated in physical terms, forming a capital stock index from the number of coal-cutting machines taking account of the improvement in these machines over time. Details of the construction of this index are to be found in appendix 6.3.

In addition, estimation was not of the production function in the form given above. Rather, in conditions of excess capacity and unemployment labour demand was taken to be the dependent variable. In other words, the question asked was: given the number of machines situated on a given number of mines what labour would be required to produce a given level of output? The final modification was to take account of capacity utilization. If output were extremely low so that mines and machines were not fully used, then the labour required to produce a given level of output with those other inputs would appear to be greater than it in fact was. As a result, use was made of the rate of growth of output as an index of capacity utilization. When output has grown, a fuller capacity utilization will reduce the labour required to produce a given output and vice versa for a fall in output.

To summarize, the equation estimated was $L = A\, Q^a\, K^b\, N^c\, G^d$ where G is one plus the rate of growth of output and d is hypothesized to be negative if excess capacity is significant. Estimates were made for each district by employing annual data from 1921 to 1938. Leaving aside G, inverting this equation yields the following relations:

$$\beta = 1/a$$
$$\alpha = -b/a$$
$$\gamma = -c/a$$

It is anticipated that $a > 0$, a positive relation between output and labour demand, that $b > 0$ since machinery reduces labour demand, and that $c > 0$ for economies of scale so that labour demand increases with the number of mines. That this last result does not depend upon worsening mine conditions can be gauged by whether $\alpha + \beta$ or $\dfrac{1 - b}{a}$ is greater than 1.

Details of the results of the regressions are given in appendix 6.2. They find that there are considerable economies of scale for most districts and that this finding is not produced by the necessity of moving on to poorer working conditions. It is found that excess capacity was an important factor in most districts in reducing productivity levels. Here is presented an average of the results, a sort of representative production function, for mining for Britain. It is calculated by taking an average over the districts, weighting by the inverse of the variance of each coefficient. The rationale and details for this are given in appendix 6.2. The production function 'on average' is

$$Q = A L^{3.6} K^{0.6} N^{-2.5} G^{0.5}$$

where A is a constant.

If anything the returns to scale are outrageously high, particularly the returns to labour. This is perhaps explained by the use of purely capital augmenting technical progress – as if only machine cutting improved in fact – whereas other forms of technical progress tending to reduce labour inputs would be ignored biasing the exponent on L upwards. It proved impossible to estimate the effects of technical progress over time separately because of the time trend on the other variables concerned.

Also, no account is taken of changes in the length of the maximum working day nor of shifts in the number of days of the year and the number of hours of the day actually worked. Further refinements might require attention to the other types of capital employed apart from the mechanization of coal cutting. Nevertheless, within the considerable limits of the exercise as a whole, it seems plausible to conclude that the coefficients indicating increasing returns may be too large but not so much so that they are qualitatively incorrect. Certainly, there must be a predisposition towards rejecting the Buxton hypothesis. As an alternative basis for future research, the production function approach might be embraced with a greater degree of sophistication. Or, preferably, the presence of economies of scale, and obstacles to their adoption, might be explored through detailed case studies (which would appear to preclude aggregate statistical techniques).

Concluding remarks

In this chapter, production functions have been estimated for the British coal industry. The primary concern has been to correct and reject the hypothesis put forward by Buxton – that economies of scale did not exist and lack of amalgamations and concentration of production was not, consequently, a serious problem. Certainly, no great significance can be placed on the estimates themselves: that output would have been much greater by so much if this that or the other factor included in the model had been different. Such counterfactual exercises do have their uses but they can also be abuses.

For it is by no means being claimed that the model employed here is an entirely appropriate one. Too many other influences of importance have been excluded and it operates at too great a level of aggregation. Moreover, only in a limited way does it make sense to assign lower levels of output for given inputs to demand deficiency or small mine size, once it is recognized that modifications of these variables would have required a reorganization of the industry in a profound structural sense – this in turn depending upon a particular resolution of class conflicts rather than a simply conceived greater or lesser improvement in the workings of the market and/or entrepreneurial efficiency.

This still leaves unanswered the problem of the industry's poor performance, as most frequently acknowledged in terms of the low pace of productivity increase and mechanization and, as those other than Buxton might have it, failure to rationalize production on to fewer mines and companies. Some twenty years ago, Paull (1968), in a comparison with the US industry, identified four possible reasons for the failure of the British interwar coal industry. First, there is the failure to rationalize and the argument is similar to that put forward by Kirby which is found to be wanting in chapter 2.

Second, the state of industrial relations lends the labour movement open to the accusation of obstructing mechanization and rationalization in its attempts to preserve existing work patterns and levels of employment. But, following the defeat in the General Strike, the miners' union was hardly in a position of strength and was further weakened in the 1930s by large-scale unemployment. Even if it had had a policy of opposition to mechanization and amalgamation, its ability to sustain it effectively must be doubted – at least in the absence of other forces pushing in the same direction.

The lack of finance is a third factor that has been seen as an impediment, particularly given the low level of profitability. But finance was made available for new sinkings in South Yorkshire and for the reorganization of the South Wales coalfield under Powell Duffryn as discussed by Boyns (1987). In addition, the argument becomes circular, with low profitability and low access to capital reproducing each other through the failure to rationalize. But such a cycle could equally be conducive to mechanization and concentration through the pooling of finance within the industry, this further intensifying competition, as Paull illustrates for the US industry, as the rationalized producers place pressure on the inefficient.

Fourth, failure to rationalize can always be explained away by natural conditions, either the coal seams or the machines (depending on the way you look at it) proving unsuitable. This does not appear to be so for the British case since mines were mechanized but only slowly. Further, the seams worked cannot be taken for granted. If unsuitable for mechanization, why was production not concentrated upon those that were suitable?

A fifth explanation for failure has been put forward by Dintenfass (1985). By a close study of four collieries, he sets aside other explanations of poor performance and shows how marketing through coal preparation was a decisive advantage for certain collieries. But it must be doubtful whether this can be generalized into an overall explanation of failure. For, suppose all collieries did invest adequately in coal preparation plant, then it must still be questioned whether this would have solved the industry's problems of low mechanization and amalgamation. In the world of the unclean, those who wash have an advantage, but this does not elevate them to the status of sartorial elegance. Dintenfass' contribution then must be seen in the context of an industry whose competitivity was not determined primarily by lowest cost of production. Why were high productivity, large-scale mechanized mines not competing with each other, with or without coal preparation?

In the light of these sorts of considerations, Paull draws the conclusion that as none of these explanations is sufficiently weighty, ultimate responsibility for only introducing change slowly must rest with the entrepreneurs of the industry. It becomes a case of residual entrepreneurial failure whenever an explanation is otherwise not forthcoming.

This approach is criticized in chapter 7 as well as elsewhere earlier. Instead, an alternative causal factor has been given some prominence, that of the obstructive role of the royalty system. In his assessment of the role of royalties, Supple's official history concludes:

> given the other reasons for withholding investment, it seems doubtful whether the royalty system prevented amalgamations (and slowed down mechanization) simply because mineowners feared that any improvement in productivity would be appropriated by the royalty owner. But the royalty system clearly helped explain the original fragmentation of the industry, presented numerous practical obstacles in structural change, and in 1933 was offered by the MAGB as the reason why the efforts of the CRMC to secure widespread mergers was doomed.
>
> (1987: 406)

This suggests an important role indeed for the royalty system, especially in the creation of the industry's industrial structure, even if it were in conjunction with other factors. It surely follows that it would equally serve to obstruct the restructuring of the industry, although this might be masked and complemented by the other problems. These in turn, however, cannot be simply read off, along with the royalties, as independent causal factors. Rather, as Supple concedes, the structure of the industry was heavily influenced by the royalty system, and it is the symbiotic relation between it and other causal factors that have to be considered as a whole. Just as it cannot be argued that the General Strike took place because of the royalty system, so the same applies to the royalty's effect on poor economic performance. But this does

Table 6.1.1 Regression estimates

District	α	$\beta 10^4$	F-statistic
1. Northumberland	91.7 (4.4)	10.5 (2.7)	4.9
2. Durham	204.4 (6.1)	10.6 (2.6)	4.7
3. Cumberland	90.0 (5.8)	5.4 (1.4)	1.9
4. Lancs and Cheshire	88.3 (8.2)	11.9 (5.2)	9.4
5. S. Yorkshire	114.6 (7.9)	4.0 (3.1)	6.0
6. W. Yorkshire	114.5 (2.3)	13.8 (2.2)	3.7
7. Notts	150.0 (12.8)	4.2 (3.5)	6.7
8. Derbyshire	79.5 (7.4)	10.9 (3.9)	7.6
9. N. Staffs	154.4 (13.2)	5.4 (2.5)	4.6
10. Cannock Chase, Warks, S. Staffs, Worcs, and Leics	58.4 (0.8)	5.9 (0.7)	0.5
11. Shropshire	126.1 (3.8)	66.8 (2.5)	4.6
12. Kent, Bristol, Forest of Dean, Somerset	−280.5 (1.5)	21.2 (4.9)	9.1
13. S. Wales	216.0 (4.0)	17.2 (2.7)	5.1
14. N. Wales	223.6 (8.2)	−2.9 (0.9)	0.8
15. Fife and Clackmannan	201.6 (3.1)	3.7 (0.9)	0.8
16. Lothians, Stirling, Renfrew, Dumbarton	113.6 (2.4)	16.9 (5.8)	10.1
17. Lanarks, Linlithglow	333.5 (8.1)	24.1 (2.2)	3.7
18. Ayrshire, Dumfries, Argyll	87.9 (2.2)	19.9 (3.2)	6.1

Note: Districts 8, 10, and 12 were composed by aggregating, respectively, two, four and four distinct districts. Of the eighteen districts, those with a small percentage of total British output were districts 3, 9, 11, 12, 14, 16, and 18 with 0.7, 3.1, 0.3, 1.7, 1.2, 2.0, and 2.0 per cent of output, respectively, in 1938. (There is little change in these proportions over the period.) Either the aggregation across districts or a small district might be a source of peculiar characteristics and there is indeed a tendency for insignificant or wrongly signed coefficients in the districts concerned.

not make it of negligible importance in either case.

Supple also tends to confine royalties to practical problems (as well as the structural and ideological as the mine owners' excuse for backwardness). As has been argued in chapter 4, that this could have occurred without economic motives is doubtful. Hopefully these can be studied in greater detail in the future along with finance, industrial relations, marketing, etc., which have tended to be more the focus of research for those breaking with the hypotheses favoured by Buxton and Kirby.

Appendix 6.1: Regressions for mechanization and mine site

Each of the following three equations was estimated for eighteen districts over the period 1921 to 1938:

$$Q/L = \alpha_i \, Q_M/Q + \gamma_i \, Q/N + \gamma_i + u \qquad (1)$$

$$Q/L = \alpha_i \, Q_M/Q + \gamma_i + v \qquad (2)$$

$$Q/L = \gamma_i \, Q/N + \gamma_i + w \qquad (3)$$

Dummies were employed for 1921 and 1926, years in which strikes substantially reduced output, and these were highly significant in general. In all but three districts equation (2) provided a better fit than equation (3) and so a test was made of the hypothesis that $\beta = 0$ in equation (1) against equation (2). In table 6.1.1, the F-statistic concerned is presented together with the estimated parameter values and their t-ratios. The hypothesis that $\beta = 0$ is rejected with 95 per cent confidence if the F-statistic exceeds 4.67.

Appendix 6.2: Regressions for economies of scale

The basic equation to be estimated for each district is

$$\ln L = a_i \ln Q + b_i \ln K + c_i \ln N + d_i \ln G + u \qquad (1)$$

On doing this it was found that the coefficients were in general significant and of the anticipated sign. However, the inverse relationship over time between capital stock and the number of mines, the first of which increased whilst the second declined, yielded high positive correlation between each b and c since they are of opposite signs. b appeared to be biased towards zero and c away from zero. On inspection, it was found that where this correlation was less severe the values of a and c across districts tended to be fairly uniform. So weighted averages were calculated of a and c across districts using the inverse of the variance of each coefficient as a weight.[8] This was done for two different assumptions underlying the formation of the capital stock index (see appendix 6.3). For a rate of technical progress in machines of 4 per cent p.a. the average values of a and c were 0.28 and 0.70 respectively and for a rate of technical progress in machines of 6 per cent p.a. the values were 0.24 and 0.60. Consequently, it can be argued that there were substantial economies of scale in the use of labour as well as for reductions in the number of mines.

With the averaged values of a and c the following regressions were estimated district by district.

$$\ln L - a \ln Q - c \ln N = b_i K + d_i \ln G + e_i + u \qquad (2)$$
$$\ln L - a \ln Q \qquad\qquad = b_i K + c_i \ln N + d_i \ln G + e_i + v \qquad (3)$$
$$\ln L - c \ln N \qquad\qquad = a_i Q + b_i \ln K + d_i \ln G + e_i + w \qquad (4)$$

The difference between the equations reflects hypotheses concerning whether a and/or c are uniform across districts. A rough guide to choosing between the equations is provided by a comparison of standard errors. In general, equation (4) performed better than the other two with little to choose between these, but overall the standard errors between the equations did not differ greatly. In addition, as before for equation (1), equation (4) introduced correlation between the coefficients on K and N. Consequently, the estimated coefficients were less uniform in sign and also exhibited fewer cases of significance. In particular, the other equations were more sensitive in picking out the 'peculiar' district. (See note to table 6.1.1 in appendix 6.1.)

The results of the estimates for equation (2) are given in table 6.2.1. They are made on the basis of an increaase in machine productivity of 4 per cent p.a. As indicated earlier, if this is amended to 6 per cent p.a. the weighted averages of a and c adjust in the expected direction. The index of capital increases explaining more of the demand for labour at the expense of mines and output. The change also had the effect of reducing the calculated values of b_i. Again, this is to be expected as more of the demand for the labour is explained by technical progress at the expense of the number of machines.

The recalculation of the regressions on the basis of a different rate of technical progress served two purposes. First, it allowed the sensitivity of the results to be assessed in relation to the techniques of estimation used, in particular for the formation of the uniform coefficients for output and mines across districts. The qualitative conclusions were robust although the parameter estimates changed in the anticipated directions already indicated, another healthy sign. No great faith can be placed on particular parameter estimates, but the uniformity of the results is striking. A more satisfactory approach would involve a district by district

Table 6.2.1 Regression estimates

District	c		d	
1.	−0.145		−0.156	
2.	−0.236		−0.102	
3.	−0.254		−0.054	(0.57)
4.	−0.266		−0.118	
5.	−0.123		−0.157	
6.	−0.224		−0.176	
7.	−0.383		−0.168	
8.	−0.158		−0.260	
9.	−0.456		−0.281	
10.	+0.090		−0.149	(1.45)
11.	−0.166		−0.046	(0.29)
12.	+0.148		−0.025	(0.38)
13.	−0.236		−0.101	
14.	−0.026	(0.17)	−0.146	(0.96)
15.	−0.054	(0.77)	−0.118	
16.	−0.131		−0.141	
17.	−0.965		−0.082	(0.72)
18.	−0.008	(0.36)	−0.096	

Note: + t-ratios given when less than 1.96.

examination of the way in which mines and machinery were introduced in response to output changes. Second, the rate of technical progress was calculated as an average over all districts but was different by as much as three times between districts.[9] Consequently, the two sets of results could be used to see whether there was improvement in 'goodness of fit' when the districts were divided into high and low productivity increasers. This was in general the case.

Appendix 6.3: A machine index for technical progress

During the interwar period there were four types of machine in use for coal cutting; bar, disc, chain, and percussive. At the beginning of the period only 37.1 per cent of coal was cut in Britain by chain machine, but this had risen to 94.9 per cent by 1938. In constructing an index of the capital stock, account had to be taken of the different productivity of machines, i.e. the service they provide, and this itself changes over time. The view was taken that the index could be constructed on the basis that machines provide services that are not district specific, but that differences in machine productivity are to be explained by other factors such as inputs and mining conditions, this last factor being captured by the constant term in a regression for a district. Suitability for machinery and suitability for mining are not identical properties but heroic assumptions are necessary for the purposes in hand.

The coal question

In order to calculate these machine services, a simple average of machine productivity (output per machine) was taken across districts for each machine. However, it was observed that machine productivity in any district in any year was highly dependent upon the level of output. Consequently, the period 1921 to 1938 was divided into three, 1921–8, 1928–33 and 1933–8. For each period and each machine, the peak productivity was calculated, then the average was formed across districts. The results are shown in table 6.3.1.

Table 6.3.1 Mean peak productivity in thousands of tons per machine

	Bar	Disc	Chain	Percussive
1922–8	12.5	12.0	13.0	3.1
1928–33	12.4	13.0	17.0	2.8
1933–8	11.7	9.8	23.3	2.7

Source: Annual Reports of HM Chief Inspector of Mines

On the basis of this information are drawn the following rough conclusions. The productivity of bar, disc, and percussive machinery remained relatively constant over time and the decline in their productivity is likely to be explained by their persistence on mines being worked out or their displacement from better mines by chain machines. Moreover, bar, disc, and chain machines were more or less of equal productivity at the beginning of the period when a percussive machine was 'worth' one-quarter of one of these machines. Chain machines on the other hand enjoyed a productivity increase of about 100 per cent. Accordingly, a doubling of productivity over the period was fitted to an exponential trend in productivity growth yielding a rate of 4 per cent p.a. over eighteen years. If, however, the productivity increase is fitted between the middle of the first and last periods, the rate of productivity increase is calculated at 6 per cent p.a. The reasons for using the different rates of technical progress are explained in appendix 6.2. Denote the rate by m; the capital stock index is now constructed in terms of 1921 chain machine equivalents by the following formula:

$$K = e^{mt} K_1 + K_2 + K_3 + K_4/4$$

where K_1, K_2, K_3 and K_4 are, respectively, the quantity of chain, bar, disc, and percussive machines in use, where m takes on the value 0.04 or 0.06.

Chapter seven

The diffusion of mechanical coal cutting

Study of the British interwar coal industry has been dominated by two propositions. The first, associated with Buxton, argues that economies of scale were not important at that time so that slow progress in amalgamation in the industry cannot be seen as a deficiency. The second, associated with Kirby, considers that 1930 legislation to organize a state cartel in the industry had the effect of cushioning the fragmented industry against economic pressures to reorganize and undermined simultaneous legislation to encourage amalgamations.

In previous chapters, it has been shown on both theoretical and empirical grounds that these two propositions are unfounded. More careful and wider econometric study of the evidence suggests that both economies of scale and mechanization were important to the growth of productivity. A study of the state-organized cartel in operation and of the pattern of amalgamations reveals that the cartel was far from effective and there is little support for the view of a slowdown in amalgamations in the 1930s.

Quite clearly, these matters are linked to the approach (and hypothesis of failure or not) of analysing entrepreneurial performance. The purpose of this chapter is to shed light on this approach by examining the process of adoption of coal cutting technology in the interwar period.[1] The evidence is mixed. From the point of view of the adoption of the best coal cutting technology, entrepreneurs can be seen to have performed well. But in adopting it at all in terms of the proportion of coal cut and especially in the number of mines mechanized, entrepreneurial performance appears to have been deficient.

In the first section, there is an informal empirical description of the introduction of mechanical cutting by looking at the amount of coal machine cut, type of machine used, and number of mines mechanized. This is followed in the second section by the study of such mechanization as a diffusion process in examining the country as a whole. This exercise is repeated in the following section for individual districts. The concluding remarks point again to the limitations of an economic

Mechanization: an overall picture

In analysing the mechanization of coal cutting it is possible to distinguish three separate processes. The first is simply the displacement of hand cutting by machine cutting and may be measured by the ratio $Q_m : Q$ where Q_m is the quantity of coal cut mechanically and Q is total output. In table 7.1 is shown the change in this ratio by eighteen districts between 1922 and 1938.[2] The second process is the movement to the best form of machine cutting, the use of the most efficient machine. For the British case, it is fortunate that a uniquely best machine can be quite clearly identified from the four types of machine that were available.[3] This is the chain machine as opposed to the disc, bar, and percussive machines. Table 7.2 shows the relative importance of these machines in coal cutting for the years 1922 and 1938. Because chain cutting had become so predominant by 1938 it is not necessary to include the percentages for other types of machine for that year, since they are negligible.

Table 7.1 Percentage of output cut by machine

		1922	1938
1.	Northumberland	20.6	91.2
2.	Durham	11.9	42.0
3.	Cumberland and Westmorland	7.6	72.9
4.	Lancashire and Cheshire	13.3	67.6
5.	Yorkshire South	9.0	56.3
6.	Yorkshire West	19.2	54.9
7.	Nottinghamshire	14.5	70.1
8.	Derbyshire	14.2	88.0
9.	North Staffordshire	21.3	92.2
10.	Cannock Chase, South Staffordshire, Worcestershire, Leicestershire, and Warwickshire	11.7	63.3
11.	Shropshire	6.5	55.5
12.	Forest of Dean, Somerset, Bristol, and Kent	2.6	14.0
13.	South Wales and Monmouthshire	3.6	26.0
14.	North Wales	15.3	69.4
15.	Fife, Clackmannon, Kinross, and Sutherland	43.6	89.9
16.	Lothians (Mid and East)	25.2	62.2
17.	Lanarkshire, Linlithgow, Stirling, Renfrew, and Dumbarton	47.1	83.9
18.	Ayrshire, Dumfries, and Argyll	18.4	69.4
Great Britain		15.3	59.4

Source: Annual Reports of HM Chief Inspector of Mines

Table 7.2 Percentage of mechanically cut coal cut by type of machine

District	1922 Percussive	Bar	Disc	Chain	1938 Chain
1.	41.8	6.4	27.2	24.6	96.9
2.	51.4	9.9	5.2	33.5	88.0
3.	6.1	39.7	0	54.2	96.7
4.	31.5	22.0	25.8	20.7	89.1
5.	7.9	22.2	13.0	56.9	97.3
6.	5.1	4.2	53.5	37.1	92.3
7.	0.3	19.5	15.6	64.6	97.1
8.	3.8	15.3	14.1	69.8	98.9
9	11.0	8.1	3.6	77.3	97.7
10.	0.8	4.4	16.2	78.6	99.5
11.	0	0	36.8	63.2	99.5
12.	37.8	47.6	19.8	4.8	99.7
13.	3.0	0.3	23.0	73.6	98.3
14.	13.8	35.5	11.7	38.9	90.0
15.	0.2	21.1	55.1	23.6	97.3
16.	0.2	46.2	31.0	22.6	90.3
17.	0.3	22.8	62.3	14.7	88.6
18.	1.5	32.0	42.9	23.6	95.5
Great Britain	12.5	17.9	32.5	37.1	94.9

Source: Annual Reports of HM Chief Inspector of Mines

The third process involved in mechanizing coal cutting is simply the movement from unmechanized to mechanized mines. For the former, there is no coal cutting machinery installed, whereas for the latter there may be a combination of hand and mechanical cutting. Between 1922 and 1938 the total number of mines in operation declined from 2,881 to 2,105 whilst the number of mechanized mines increased from 785 to 927. Thus, whilst there was an absolute increase in the number of mechanized mines in a declining total of mines, the percentage of mechanized mines remained below 50 per cent. Table 7.3 gives the district by district movement of these variables between 1922 and 1938.

This overall picture of the process of mechanization leads to some empirical observations. The introduction of the best technology, namely cutting by chain machine, proceeded at a much greater pace than mechanization itself. This suggests that entrepreneurs did not fail in the sense of failing to choose the right technology when they chose to mechanize. If it is supposed that entrepreneurs who can make the correct choice between the technologies of mechanization are also liable to be able to make the correct choice between mechanization or not, then the evidence supports the conclusion that it is unlikely that individual entrepreneurial failure can explain the slow pace of mechanization. The

Table 7.3 Mechanization by mine

District	Number of mines				Percentage of mechanized	
	Mechanized		Unmechanized			
	1922	1938	1922	1938	1922	1938
1.	43	66	84	36	33.9	64.7
2.	83	93	179	154	31.7	37.7
3.	6	12	32	13	15.8	48.0
4.	87	80	180	75	32.6	51.6
5.	38	67	100	54	27.5	55.4
6.	53	48	137	68	27.9	41.4
7.	17	41	22	5	43.6	89.1
8.	31	59	115	57	21.2	50.9
9.	29	29	59	31	32.9	48.3
10.	37	48	202	95	15.5	33.6
11.	3	4	56	43	5.1	8.5
12.	8	14	78	35	9.3	28.6
13.	77	118	565	306	12.0	27.8
14.	17	10	30	18	38.2	35.7
15.	49	44	19	7	72.1	86.3
16.	21	25	20	8	51.2	75.8
17.	165	132	148	152	52.7	46.5
18.	21	37	70	21	23.1	63.8
Great Britain	785	927	2096	1178	27.3	44.0

Source: Annual Reports of HM Chief Inspector of Mines

limited movement from unmechanized to mechanized mines adds further evidence to the view that there were constraints on mechanization at an aggregate level rather than a failure of individual entrepreneurial initiative, for it is implausible that entrepreneurs could have the wit to mechanize in the right way when they did do so but otherwise fail to mechanize at all or insufficiently intensively within mechanized mines. Even if this is explained by one set of innovative mine owners and another of dim-witted mine owners, why the first did not competitively eliminate the second remains a mystery. Before returning to this issue, a closer look will be taken at the detailed empirical results.

The diffusion of mechanization in Britain as a whole

Begin by treating the development of mechanization as if it were a diffusion process, not necessarily because it is plausible that entrepreneurs, learning about and introducing mechanization, shared a cumulative experience akin to the latest craze in fashion or pop music. Rather it acts as a means of describing the introduction of a new process over time. It does so by the use of three parameters; one is the level of mechanization at the beginning of the period considered, the second is the speed with which mechanization takes place, and the third is the ultimate level of mechanization to which the diffusion is tending, a level which may well be estimated to lie below 100 per cent. If p represents the percentage of mechanization, then a diffusion process is illustrated by the graph in Figure 7.1.

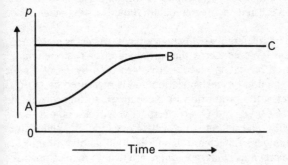

Figure 7.1 A diffusion process

The parameters referred to above are related to A the starting point, the steepness of the gradient AB giving the speed of adjustment, and C yielding the limit of mechanization.[4] Concern will focus on the last two parameters since the starting level is approximately given by the actual levels of mechanization at the beginning of the period and these are presented in the tables in the previous section.

This section will examine the diffusion of mechanization for Britain as a whole. Before doing so, it is worth emphasizing that the movement of a measure of mechanization is only being described as a diffusion process for convenience. In doing this, there are three possible outcomes. The first is that a movement may not be according to a diffusion pattern since the level of mechanization may decline, stagnate, or fluctuate rather than increase. In this case, the result of describing it as a diffusion process will lead to insignificant or wrongly signed estimates of the parameters. Second, a diffusion pattern may be found to be significant with a limiting tendency of unity (i.e. 100 per cent) together with an estimated speed of adjustment. Finally, a diffusion pattern may be significant with a limiting tendency less than unity that can be estimated together with an associated speed of adjustment. If the movements in a particular ratio were estimated for the last two cases, then irrespective of the better of the two fits, the first estimate is bound to yield a lower speed of adjustment than the second. For the pattern of adjustment would be interpreted as either a relatively slow movement to unity or a relatively fast movement to a lower limiting ratio.

In appendix 7.1, it will be shown that there are two ways of testing whether mechanization has a limiting tendency of unity or not.[5] After this test, in each case, the adjustment speed can be calculated and the overall significance of the adjustment as a diffusion process assessed. In this section are presented the results of these two cases for Britain as a whole and, in the following section, the results for a district by district analysis. Those interested in greater detail should at this stage read appendix 7.1.

The discussion has differentiated between three processes associated with mechanization. One concerns the introduction of best technique given mechanization has taken place and is measured by Q_c/Q_m; one concerns the introduction of mechanization and is measured by Q_m/Q; and the last concerns the proportion of mechanized to unmechanized mines, N_m/N. Here Q, Q_m and Q_c are, respectively, total output, output cut mechanically, and output cut by chain machine. N_m and N are respectively the number of mechanized and the total number of mines. As indicated in the previous paragraph, there are two ways of estimating diffusion. The results are presented by use of the first method for each form of mechanization in table 7.4a and for the second method in table 7.4b. In each case, the estimates are significant, so that the movements can be described as a diffusion.

Table 7.4 Diffusion of mechanization in Britain

	(a)		(b)	
	Limiting Tendency	Speed of Adjustment	Limiting Tendency	Speed of Adjustment
Q_c/Q_m	1	0.215	1	0.183
Q_m/Q	1	0.141	1	0.116
N_m/N	1	0.045	0.42	0.201

What the results reveal is that the pace of mechanization was faster for the adoption of best technique than for mechanization as a whole which was in turn faster than the transformation of unmechanized to mechanized mines. Although according to table 7.4b the speed of adjustment was estimated to be higher than the others for the mechanization of mines, this resulted from the fitting of a diffusion process to an extremely low level of ultimate mechanization, 42 per cent.[6]

Mechanization district by district

The three processes of mechanization can also be described as diffusions at a district by district level. Of the two methods of estimating the parameters concerned, one alone was eventually chosen for its preferable statistical properties. In general it was found that for both the ratios Q_c/Q_m and Q_m/Q the limiting tendency of mechanization was unity. This is perhaps not surprising for Q_c/Q_m given its value of around 90 per cent or above for most districts by 1938, but a little more surprising perhaps for the second ratio given the relatively low level of mechanization in some districts by the end of the period. Table 7.5 gives the estimates of the speeds of adjustment for the ratios when a diffusion process is fitted for each and tending to a limit of unity. In general (eleven of eighteen districts), in uniformity with the results for Britain as a whole, it is found that the speeds of adjustment for Q_c/Q_m are greater than those for Q_m/Q. Where this is not the case, it can be explained by the initial high proportion of chain machinery in mechanization but low proportions of mechanization as a whole. In other words, the broad conclusion to be drawn is that the adoption of best coal-cutting technology was done more rapidly than the adoption of mechanization as such.

Table 7.5 Speeds of adjustment for mechanization

District	Q_c/Q_m	Q_m/Q
1.	0.333	0.234
2.	0.160	0.098
3.	0.165	0.304
4.	0.239	0.203
5.	0.163	0.219
6.	0.155	0.102
7.	0.157	0.200
8.	0.228	0.253
9.	0.126	0.266
10.	0.232	0.166
11.	0.135	0.185
12.	1.431	0.118
13.	0.0743	0.158
14.	0.239	0.140
15.	0.257	0.136
16.	0.269	0.0958
17.	0.232	0.120
18.	0.260	0.150

The results of fitting diffusion processes to the mechanization of mines produced completely different results from those for Q_c/Q_m and Q_m/Q. For six districts, the mechanization of mines could not be viewed as a diffusion process at all. For seven districts, it was best described as a painfully slow adjustment to 100 per cent mechanization. For the remaining five districts, four adjusted quickly to very low levels of ultimate mechanization and only the last adjusted quickly to a high level of mechanization. These results confirm the picture drawn at an aggregate level that the mechanization of mines lagged behind the mechanization of coal cutting.

Concluding remarks

One response to the sort of evidence considered here is to situate it within the hypothesis of entrepreneurial behaviour. Decisions to adopt new technology or not are made on the basis of the comparative advantage of mechanized and unmechanized mines by reference to their estimated costs of production given relative prices of factor inputs.

This is the method adopted by Greasley (1979 and 1982). He estimates the optimum level of mechanization and a diffusion process towards it, looking at the different working conditions across districts and whether they are more or less favourable to the adoption of machine methods. Greasley (1979: 267) concludes that 'statistical support was found in the majority of coalfields for the view that the equilibrium demand for machine use shifted through time in response to changes in technology and the economic environment, but that the actual use of the machine technology only adjusted slowly to equilibrium'. Leaving aside the question of whether there was or could be a fixed equilibrium towards which the industry was moving, this conclusion concerning the pace of movement is in agreement with the results presented here.

However, the analysis involved cannot be accepted even if the conclusions are found to be palatable. There are the problems associated with Greasley's use of economic aggregates at current prices to equate marginal products to factor rewards and the dependence upon the full employment of resources and constant returns to scale. These are necessary to assess the relative benefits of capital- and labour-intensive techniques. Even so, even if the relative pace of mechanization between districts could be explained by some combination of relative prices and geological differences, this would not explain the overall pattern of mechanization within districts according to the three processes identified here, particularly the persistence of unmechanized, presumably small-scale, mines. The only possibility is that within each district there is the co-existence of one group of entrepreneurs who are 'rational' and another 'irrational', from the perspective of optimization, and that the one does not eliminate the other through competition. This seems implausible and would, in any case, demand an explanation of what could produce, let alone sustain, such a bi-modal distribution of employers.

An alternative explanation has been offered by Buxton (1970) by drawing a distinction between the relative advantage of new and old investment. For new investment to be justified, its total average cost of production must be below the variable cost of production of existing capacity for which investment expenditure has already been sunk. This clearly has some relevance for a period in which there was considerable excess capacity, supporting the idea that existing capacity efficiently crowded out better methods.

It must be doubted, however, whether the picture is accurate, of old mines, as they were exhausted, gradually giving way to the new without in the meantime receiving any further investment which could have reasonably included mechanical coal cutters. Moreover, this analysis remains at the level of the static optimizing behaviour of the individual entrepreneur. A more realistic picture might be the one, in the context of mass unemployment, in which the introduction of labour-saving technology swells the ranks of those miners open to exploitation in the less progressive mines. In other words, the process of mechanization could help to sustain the persistence of the unmechanized.

Before leaving the subject of entrepreneurial behaviour, it is worth observing that the whole issue of individual rationality is thrown into doubt once the questions of collective action and interests are introduced.[7] Certainly, the mine owners were capable of behaving individualistically, refusing to accept national wage bargaining (although negotiating collectively at a district level), and in impeding schemes to reorganize the industry in the 1930s. As such, they could paradoxically be thought of as collectively defending the practice of individualism! Equally, however, the owners welcomed the state-organized scheme in the 1930s to support coal prices.

Thus, the extent of individual and collective action has to be determined, if it is accepted that the distinction between the two is fundamental, whereas hypotheses around entrepreneurial behaviour tend to take the arena of collective action as given (and usually negligible as in measurement of total factor productivity with its sole reliance upon individual optimization in conditions of perfect competition). In short, where there is, as always, (potential) collective action and organization of employers, the notion of individual rationality (equals optimization) becomes ill-specified. This can be accommodated either within game theory or within a notion of collective action to breach the (institutional) constraints within which individuals operate. Either way this is a much needed step towards the understanding of the economic and social relations within which capital and labour (fairly basic collectivities) are organized and develop, even if it is only a small part of the journey in departure from the neoclassical orthodoxy.

The coal question

Appendix 7.1 The econometric estimates

Three different processes associated with mechanization have been distinguished. For the moment denote each of the ratios Q_c/Q_m, Q_m/Q, and N_m/N as p in order to obtain the appropriate mathematical form of a diffusion process.[8] The standard form of adjustment for p is given by:

$$\frac{ap}{1-ap} = be^{ct} \qquad (1)$$

where by rearranging:

$$p = \frac{1}{a}\left\{\frac{be^{ct}}{1+be^{ct}}\right\} \qquad (2)$$

a can be interpreted as the limit of the level of mechanization to which the system is tending; c is the pace of mechanization; and b is related to the initial level of mechanization.

By taking logs of equation (1):

$$\log\frac{ap}{1-ap} = \log b + ct + u_t \qquad (3)$$

where u is an added error term. This equation can be estimated by selecting a value of a and regressing $\log[ap/(1-ap)]$ on t. This can be taken further by estimating the equation for a selection of values of a and then choosing the one that gives the best fit for (3). An alternative procedure is to differentiate (2) and obtain equations to estimate in the form of first differences. There are two natural ways to do this. From (2):

$$\frac{1}{p} = a\frac{(1+e^{-ct})}{b}$$

$$\frac{d(1/p)}{dt} = -\frac{ace^{-ct}}{b}$$

$$= -ac\left(\frac{1}{ap} - 1\right) \qquad \text{from} \quad (1)$$

$$\frac{d(1/p)}{dt} = ca - c\frac{1}{p}$$

which can be estimated in discrete form as:

$$\frac{1}{p_{t+1}} - \frac{1}{p_t} = \alpha + \beta \frac{1}{p_t} + v_t \qquad (4)$$

where $a = \alpha/\beta$ and $c = -\beta$.

Alternatively, from (2):

$$p = \frac{1}{a}\left\{1 - \frac{1}{1+be^{ct}}\right\}$$

$$\frac{dp}{dt} = \frac{bce^{ct}}{a(1+be^{ct})^2}$$

$$= \frac{c\dfrac{ap}{1-ap}}{a\left(1+\dfrac{ap}{1-ap}\right)^2} \qquad \text{from} \quad (1)$$

Simplifying, $\dfrac{dp}{dt} = cp - acp^2$

which can be estimated in discrete form as:

$$p_{t+1} - p_t = \alpha p_t + \beta p_t^2 + w_t \qquad (5)$$

where $c = a$ and $a = \beta/\alpha$.

Equations (3), (4), and (5) were each estimated over 1922–38 for the ratios Q_c/Q_m, Q_m/Q, and N_m/N. For equation (3) a was set equal to 1. Equations (4) and (5) were also estimated in the constrained forms for which $a = 1$ (i.e. $\alpha = -\beta$), respectively:

$$\frac{1}{p_{t+1}} - \frac{1}{p_t} = \alpha\left(1 - \frac{1}{p_t}\right) + v_t \qquad (6)$$

$$p_{t+1} - p_t = \alpha(p_t - p_t^2) + w_t \qquad (7)$$

The coal question

A simple F-test could then be used at the 5 per cent level to test whether a in each case was significantly different from unity. The interpretation of accepting the hypothesis that a equals one is that the level of mechanization concerned is tending over time to 100 per cent. If this hypothesis is rejected than an estimate is obtained of $1/a$ as the limit to which the process is tending when $a > 1$ or, for $a < 1$, the description of the process by diffusion is rejected. When not unity, a is calculated from (4) as $-\alpha/\beta$ and from (5) as $-\beta/\alpha$.

The results

In principle there are three sets of equations to estimate, those associated with (3), (4) and (5). Although derived from the same adjustment process (1), at most only one of these equations can have the structure of error terms appropriate for ordinary least squares estimation, for if one does this implies that the others do not. The estimation then concerns not only the calculation of parameter values but also the determination of the 'best' of the three equations in terms of the structure of errors. For reasons which will become clear, first are discussed the estimations for the ratios Q_c/Q_m and Q_m/Q and then attention is turned to the ratio N_m/N. The estimates of equation (3) for each district give significant and positive values for c in each case for both Q_c/Q_m and Q_m/Q but also rendered extremely low Durbin–Watson statistics. For Q_c/Q_m equations (4) and (6) accepted by F-tests the hypothesis that $a = 1$ for all but district 18 and equations (5) and (7) did the same for all but district 16. For Q_m/Q equations (4) and (6) accepted $a = 1$ for all districts except 11, 14, 16 and 18 and equations (5) and (7) accepted $a = 1$ for all districts except 2, 8 and 18. Given the acceptance of $a = 1$ in general, there remains the choice of estimates between those given by equation (6) and those given by equation (7). According to the Durbin–Watson statistics, both were quite good with a slight edge possibly enjoyed by equation (7). For a comparison of heteroscedasticity between the equations, a method was used as suggested by Breusch and Pagan (1980). The errors for each of the equations were regressed upon the variable p (for $p = Q_m/Q$ and Q_c/Q_m) and a comparison made of the resulting R^2 for each ratio, since the equation with the lower R^2 enjoys the lower degree of heteroscedasticity. For the ratio Q_c/Q_m the equation based on p rather than on $1/p$ had a lower R^2 in every case (on average 0.1189 as opposed to 0.2002) and for Q_m/Q did so for all but six districts (on average 0.0785 as opposed to 0.1304), these six tending to be associated with the unusual result of rejecting that the limiting proportion of mechanization was 1. The results for the estimates of equation (7) are shown in table 7.5 whereas those which rejected the hypothesis that the limiting tendency was unity are shown in table 7.1.1.

Table 7.1.1 Estimates from equation (5) rejecting limiting tendency of unity

District	Q_c/Q_m Limit $1/a$	Speed of adjustment c	Q_m/Q Limit $1/a$	Speed of adjustment c
2.	—	—	0.49	0.392
8.	—	—	0.69	0.132
16.	0.89	0.435	—	—
18.	—	—	0.71	0.400

Inspection of table 7.5 in the main text reveals that in eleven of the eighteen districts the speed of adjustment associated with the adoption of chain machinery was greater than that for mechanization. In the districts for which this was not true, namely 3, 5, 7, 8, 9, 11, and 13, tables 7.1 and 7.2 in the main text show that this can be explained by two factors; that these districts started from very low levels of mechanization as a whole but from very high levels of chain machine usage in that mechanization, with this latter factor being particularly important. Thus, the unweighted average of chain machine cutting for these districts in 1922 was 65.7 per cent compared with 37.1 per cent for the country as a whole. Necessarily the adoption of chain machinery will be seen as a relatively slow adjustment process. In addition, the unweighted mean of percentage of coal cut mechanically for these districts in 1922 was low at 11 per cent (compared to 15 per cent for the UK) whilst the average for 1938 was 65.6 per cent (72.3 per cent excluding district 13 with 26 per cent) as against 69.4 per cent for the UK. This suggests that the relatively fast adjustment for mechanization as a whole as estimated by diffusion is predominantly the result of the high level of chain usage at the beginning of the period in these districts. Thus, whilst the average speed of adjustment for mechanization for all eighteen districts is 0.175 and for these seven districts is 0.226, for chain cutting the average for all districts is 0.269 and for these seven districts is 0.150. This suggests that the faster speeds for the former are to be more explained by the slower speeds of the latter (an average reduction of 79 per cent) than by the faster average speeds of the former (an average increase of only 29 per cent). The broad conclusion that can be drawn is that the adoption of best coal-cutting technology proved far more rapid than the adoption of mechanization as such.

The results of the regressions for the ratio N_m/N are in sharp contrast for those for Q_c/Q_m and Q_m/Q. For equations (4) and (6) nine districts rejected the hypothesis that $a = 1$ and five did so for equations (5) and (7). However, more important, whether rejecting the hypothesis or not, the better equation for each of five districts in the first case and

The coal question

for each of six districts in the second yielded insignificant coefficients at the 5 per cent level. The results for equations (5) and (7) are presented in table 7.1.2.

Table 7.1.2 Movements in N_m/N

District	if F-test successful for a = 1	Speed of adjustment c	1/a if F-test fails
1.		0.081	
2.		0.613	0.38
3.		0.611	0.37
4.	Insignificant results		
5.		0.077	
6.		0.036	
7.		0.139	
8.		0.086	
9.	Insignificant results		
10.		0.349	0.35
11.		0.485	0.09
12.	Insignificant results		
13.		0.067	
14.	Insignificant results		
15.		0.937	0.83
16.	Insignificant results		
17.	Insignificant results		
18.		0.108	

Inspection of the results for the 'diffusion' of unmechanized to mechanized mines shows that this process was unambiguously slower than the other two. For those seven districts which were tending to eliminate unmechanized mines altogether (districts 1, 5, 6, 7, 8, 13, and 18), the pace of adjustment was painfully slow, 0.085 on average. For those districts 2, 3, 10, 11, and 15 for which adjustment was to less than full mechanization, the speeds of adjustment were very fast, but to low limits with the exception of district 15 (see table 7.1.2), which maintained a high level of mechanized mines throughout the period (see table 7.3). The remaining districts 4, 9, 12, 14, 16, and 17 could not be described as exhibiting a diffusion process with any statistical significance.

Part four
Towards privatization?

Chapter eight

The commanding heights of public corporation economics

Towards the end of 1988, two events in particular symbolized the extent to which the Thatcher Government had been able to make a success out of its privatization programme.[1] The first was the announcement by Cecil Parkinson, then the Energy Minister, of the proposed 'ultimate' privatization, that of the National Coal Board, which had been newly named the British Coal Corporation after the strike of 1984/5 to reflect its increasingly commercial objectives. No doubt this was purely a domestic affair, although many multinational energy companies would surely be taking an interest in the assets that were to be on offer.

The second event signalled the birth of privatization as Britain's latest export in the realm of ideas. The country that had given the world Keynesianism had, no doubt, lost the intellectual race to monetarism to the United States' Milton Friedman, but it had regained the initiative in the export of economic policy with its newest method for rolling back the frontiers of the state. Whilst this export might do little to improve the balance of trade, it promised much more in the balance of invisibles through City consultancies. For the scheme of advice was being put forward by the Overseas Development Agency (ODA), admittedly in conjunction with US agencies, to advise foreign states how best to bring about privatization. It was ample testimony to the leadership that the UK had taken in privatization, even though the latter was already a global phenomenon with the number of countries adopting such policies bordering on triple figures.[2]

In the literature that has responded to this dramatic change in industrial policy, it has been usual to point out that there are four different aspects to privatization – the selling off of assets, the contracting out of services, making charges for what was previously free, and liberalizing regulations so that, in principle, capital can flow more freely into a sector and enhance competition. In this chapter, concern is primarily with the first of these, particularly the denationalization of the major public corporation.

It has also become commonplace to list the possible motives for privatization and why they have matured. These include the ideological gains of the New Right (see Miliband *et al*. 1987), antipathy to trade unions, the wish to reduce the public sector borrowing requirement, a perceived failure in the performance of the nationalized industries, shaking up a complacent and unimaginative entrepreneurship arising out of the cosy relation between public sector management and the civil service – especially in the absence of market forces, the personal commitment of Mrs Thatcher to the *laissez faire* economy, the wish to pre-empt the ability of a future Labour Government to rescind the imposition of commercial operation on the industries concerned, the influence of the City, the wish of Ministers to unburden themselves of troublesome responsibilities, and the pursuit of popular or people's capitalism through means of wider share ownership. (See Wiltshire 1987, for example.)

With such a list of influences, and there are others potentially more appealing to the economist, such as the erosion of natural monopoly and externalities through technical change, it would surely be possible to explain almost anything. Obviously the particular causative weight and historical timing of these influences would, on the other hand, be difficult to disentangle. As a result, if the development of privatization is to be understood (and assessed), the theoretical framework for doing so becomes extremely important. Otherwise, it will be a matter of weighing up the factors that operated for and against and, inevitably, concluding that the former were stronger.

This chapter examines different theoretical frameworks within which privatization has been understood. Concern will be predominantly with denationalization; see Ascher (1987) and Forrest and Murie (1988), for example, for consideration of contracting out and selling off of council housing, respectively. Not surprisingly, the changes in policy towards the nationalized industries have resulted in a response from economic theory. In the early 1980s, the economic analysis of the public corporations was, to oversimplify, moribund and dominated by the longstanding proposition that prices of products should be set equal to long-run marginal cost and investment should earn the social discount rate – usually taken to be lower than the market rate of interest. This has now changed and two new schools have come to prominence which, quite rightly, emphasize the issue of how optimality is to be achieved (rather than what are its abstract properties in terms of the differential calculus).

The first school is the neo-Austrian which is highly supportive of *laissez faire*. It has had the most profound influence on policy making, at least in the UK. It is critically assessed in the following section. The second school, which will be termed the synthesis, has been highly critical of the Government's policies, but arguably with limited influence. It has been much more successful in establishing itself as the new orthodoxy in academic circles. It will be outlined and critically assessed in the second section. The final section examines the issue of regulation by way of conclusion.

The economics of the crank-tank

There can be little doubt that the changing balance of political power in the UK has been crucial in bringing about policies of privatization, especially in view of the long-lasting presence of the Thatcher Government. Privatization is in part a policy both to make that shift and to reap its rewards. As Pirie, President of the Adam Smith Institute, puts it in Walker (1988: 3–5) in a fictional portrayal of Mrs Thatcher's motives immediately after her first election victory:

> 'Who else tends to vote against us?'
> 'Prime Minister, we do find that people who work for the state-owned firms in the public sector tend to vote against the Conservative Party.'
> 'Right, I want less of them.' So, there are now two-thirds of a million fewer of those.

Thus, along with policies to wipe out Labour voters in council houses, trade unions and the civil service(!) and to create Tory voters through home and share ownership and the self-employed, Pirie sees privatization as a political strategy.

In this, however, it is equally important to recognize that a shift in the balance of power is involved and this partly reflects the weakness of the Labour movement's response to so-called 'Thatcherism'. This has been a weakness in campaigning, resistance, and ideological stance. As Veljanovski (1987: 24) observes:

> After seven years of Conservative Government, the Labour Party has yet to formulate a credible and intellectually defensible alternative to privatization.

The New Right itself puts this even stronger, Redwood in Letwin (1988b: 13):

> Privatization and success are increasingly identified with one another. A particular testimony to the extent of this success is the change in the British Labour Party's attitude towards privatization . . . . Between 1983 and 1987, as the scale and pace of the privatization programme increased, the Labour Party gradually abandoned this sort of rhetoric (increasingly in favour of public ownership and) . . . talked of 'social ownership'.

And this ideological shift is seen as part and parcel of a more general rolling programme in which privatization drags public opinion behind it rather than being pushed forward by it, as is also testified by the New Right (Walker 1988: 15):

> I tend to think it (privatization) was an accident . . . we didn't get public opinion to support it until after it was done.

This situation has placed the exponents and policy makers of the New Right in an advantageous position and, once this opportunity was seized, they have not been slow in exploiting their newly found potential. But, in the context of denationalization, this only seriously occurred after the re-election of the Thatcher Government in 1983 (see Brittan 1986). It seems to have been adopted rapidly to sustain what was judged to be the favourable political momentum created by council house sales which had themselves by the end of 1988 almost matched the revenue raised by denationalization. Thus, for Letwin (1988a: 66):

> Before 1983 . . . nobody had a coherent policy, nobody knew how to overcome the tremendous inertia and obstacles inside the bureaucracy . . . there was a universal dither.

Once the re-election watershed had been reached, however, the captains of privatization were prepared to sail through all possible impediments, since privatization itself had become the overriding principle of policy. Neither difficulties nor, one suspects, efficiency and rational argument, were to be allowed to stand in the way.

'Go forth and privatize' was the order of the day with sufficiently flexible *ad hoc* legislation to overcome any future objections and difficulties, Walker (1988: 14) and, for Letwin (1988a: 61):

> So, my first advice is to take the plunge. My second advice is, don't listen too much to too many experienced professionals. Choose advisers who will take the plunge with you and get on with it. You will find as I say, that it's not perfect, but it does work.

And, since 1983, the British Government certainly has taken the plunge. It had by 1988 sold off almost £20 billion worth of assets in eleven billion shares, including eleven listed companies with a total capitalization of over £58 billion.[3] Most prominent have been the sale of British Airways, British Gas, and British Telecom. Privatization of the electricity industry promises as much revenue as this all over again.

And privatization has been a big, expensive, and profitable business for UK financial institutions at home and abroad. The private sector institutions involved in arranging the sales are listed in Price-Waterhouse (1987), Dick (1987) listing some merchant bank involvement in the UK and around the world, Blanden (1988) discussing the private company advisers for a number of countries, and Brown (1987) the role of the eight large accounting firms. In general, the British privatizations have been specifically criticized for the selling off of the family silver much too cheaply, with stock market trading prices far exceeding issue prices, as laid out, for example, in Mayer and Meadowcroft (1985). If the motive has been to make wider share ownership attractive, there has only been limited success. For most small-scale punters have realized their gains by selling soon after buying, and those that have not have only held on to a few shares in a single company.

More recently, Buckland (1987) estimates that from 1979 to 1987, between £600 and £1,300 million was lost through under-pricing, with £833 million (or 4.2 per cent of proceeds) being spent on administrative and other expenses. He suggests that other methods of sale, such as repeat offers and tendering, could have raised much more revenue. France performed much better in this respect with its privatizations.

The UK's poor performance in getting the issue price has been rationalized thus, Walker (1988: 20):

> There is a mansion somewhere in South America where retired Nazi generals argue with British financiers, and the Nazi generals show that in hindsight they, in fact, won World War II seven times over. And the British financiers show that with hindsight they could have gotten a very much better price for Telecom, British Gas or whatever.

John Moore (1986: 5), at the time a Minister responsible for privatization, explains away the discrepancy between the selling price and the much higher trading price of shares by appeal to the instantaneous improvement in the industries' prospects:

> They look at the market price in the period since sale and, if it has gone up, they say that the Exchequer has lost out. What nonsense! The share price performance of companies after privatization is at least partly a reflection of how the profitability and efficiency of the companies have increased as a direct result of being privatised. At the time of privatisation, the exact scope for improvement in performance is not at all clear, and so it is not surprising that some of the companies have performed better in the private sector than was generally expected.

The truth is rather different. For the share prices acted exactly as expected due to the issue prices being too low, and this has nothing to do with the short time since privatization in which prospects could be reassessed.

Such errors might also be justified on the grounds that the privatization programme is a complex institutional readjustment requiring collaboration between industry, finance, and government.[4] However, co-operation has been promoted by the mobility between these separate arenas. Thus, John Redwood was a think-tank adviser to Mrs Thatcher. He confesses, 'On leaving Downing Street in order to take up a prospective Parliamentary candidacy I had to find a way of earning my living'. He found one advising Rothschilds in the formation of its privatization unit. This openness is to be found in the foreword to Oliver Letwin's book, *Privatising the World*. Its author was formerly with 10, Downing Street's privatization unit. No doubt practising what he preaches, he moved into the private sector to become Head at Rothschilds' International Privatization Unit. Also, Howard Hyman, for example, previously seconded to the Treasury to advise on privatization, returned to Price-Waterhouse as Director of Privatization Services.

The Government does, however, appear to take its responsibility seriously in picking stockbrokers and advisers to arrange the issues, this being described as a competition akin to the close scrutiny associated with a beauty contest, so closely are merchant banks, accountants, and consultants examined (Letwin 1988a: 136). Judging by the under-pricing, perhaps the Government is right in suggesting that it would be equally incompetent in picking industrial successes, this being no business of civil servants in view of its *laissez faire* ideology! But, as has been seen in this instance, members of the civil service, government, and the private sector are not always so far removed from each other.

No doubt such relations between the formulation and the implementation of policy, as between the public and the private sector, are not unusual,[5] although they may be specifically important in the context of the Thatcher Government in so far as she is judged to have been more than normally forceful in overcoming Civil Service and corporate management opposition to radical changes in policy. Further, the intellectual basis for privatization has been provided by right-wing think-tanks such as the Centre for Policy Studies (CPS), the Institute of Economic Affairs (IEA), and the Adam Smith Institute and, again, this has led to a blurring of the distinction between independent assessment, policy formulation, and private entrepreneurship.

So far in this section, there has been primarily a focus on what would normally be considered to be a polemical critique of the relationship between a right-wing government and its advisers. This does, however, serve as a reminder that the intellectual content of the New Right's approach to the economy does not operate in a vacuum but has been associated with a more than usually close relationship between the formulation, implementation, and propaganda of privatization. Moreover, it will be shown that the practical imperative of 'go forth and privatize' has its counterpart in the intellectual dogma of the neo-Austrian school.

In this the chief academic representative and an important policy adviser for the Government appears to have been Stephen Littlechild. He has had close links with the IEA and has shot to prominence as a result of his intellectual approach which is admitted by himself to have been on the margins of mainstream economics, referring to ideas (Littlechild 1979: 207) 'pioneered by a dozen or so other economists over the last fifteen years'. (See also Littlechild 1978, Chapter II.) These ideas do little more than to reproduce the propositions of the Austrian school of economics associated with Hayek.

The Austrian school is *laissez faire* but on different and antagonistic grounds to the outlook of monetarist macroeconomics with its neoclassical foundations. It emphasizes how the market allows individual entrepreneurial initiative to prosper and perceives this as a dynamic element of change rather than seeing the market in the static context of a full and efficient allocation of resources. Thus, Littlechild (1978: 11):

> It might be argued, therefore, that Austrian economics provides a *generalisation* and *redirection of neo-classical thought*, rather than a 'root and branch' replacement for it.

The superiority of the market is almost axiomatic and unchallengeable (Littlechild 1978: 25):[6]

> It is not surprising, therefore, to find that Austrians have for the most part eschewed empirical and statistical work. They have concentrated on deriving propositions of a *qualitative* rather than quantitative nature. These propositions follow from the basic insights into human nature . . . . For this reason, it is envisaged that such propositions will be true for *all* times and places rather than only for specific times and places.

As such the market should be made inviolable to the intrusion of special interest groups exercising pressure on government, by the creation of a long-standing legislature with slow annual replacement of its members (Littlechild 1978: 79):

> The purpose of this separation of powers is, of course, to create a legislature which is not subservient to the momentary pressures of government, and hence which severely limits the responses which governments make to immediate political pressures, in order to protect the long run interest of these same people.

In effect, private property is to be enthroned at the expense of electoral democracy so that the people can be safeguarded against the adverse effects of the pressure on government of the selfish pursuit of their own particular individual or collective interests, insufficiently moderated by the market place. No doubt regulators of privatized corporations fall into this category of legislature.

In so far as there is an innovation in this thinking it is to recognize and to draw upon the theory of property rights.[7] This tends to argue that as many aspects of the economy as possible should be owned and marketed. This, for example, would deal with abuse of the environment (Littlechild 1978: 75):

> It does, of course, raise the question whether conservationists will be able to raise sufficient funds to protect scenery, or persuade taxpayers to do so. If not this would indicate that scenic delights are a minority interest.

By this approach, the process and ability to protect the environment is very much confined to the dimensions of economic power with the result that the National Trust is held up as an example of how more general environmental interests are otherwise to be served in practice. Whatever the quality and scope of this charity's work, it and others like it can hardly be expected to take on the responsibility of negating the undesirable effects of the single-minded pursuit of profit making.

One effect of this form of support for *laissez faire* is to downplay the significance of monopolies or monopolistic behaviour. In contrast to the neoclassical's emphasis on the static deadweight loss of monopoly, the Austrian school perceives the pursuit of monopoly profits as innovative, and competition through entry to the sector concerned is the guarantee of efficiency.[8] As an example, Littlechild refers to the repayment to the NHS by Hoffman-La Roche of excess profits earnt on Librium and Valium, arguing that this will have the negative effect of stifling the development of new drugs. Given the damaging epidemic proportions in which these drugs have been used, this might not, by others, be considered such a bad thing! Elsewhere, Littlechild (1981: 31-2) has praised advertising as the bearer of information for consumers, 'there is an important role for the manufacturer in bringing these new products to their notice', setting aside the high pressure salesmanship that is used by pharmaceutical companies, for example, within the medical sphere, with the ultimate customers, patients, receiving pills, not information.

The benevolence with which monopolies are perceived is matched by an equal hostility to nationalized industries, a paradox in so far as the standard orthodox critique of public corporations is that they exploit a statutory monopoly (Littlechild 1979: 8):

The misallocation of resources is an inherent consequence of nationalisation, and in many cases the public interest could better be served by denationalising industries or 'hiving off' parts of them . . . the introduction of private capital and the abolition of restrictions on competition are essential elements of an adequate framework for controlling the remaining nationalised industries.

This, then, is the ideological background of the economist who was to be appointed to the Monopolies and Mergers Commission and, in May 1989, it was announced in Parliament that he had been chosen to be the future Director General or Regulator of the privatized Electricity Supply Industry – at an undisclosed salary and after a reputed world-wide search. He had previously advised the Government on electricity privatization (payment forming part of a £6 million bill, as disclosed by the Energy Select Committee (1987–88b)) and had written Government policy papers on the privatization of British Telecom and the Water Authorities (Littlechild 1983 and 1986 respectively). For the former, it was argued that regulation would be temporary until such time as the industry became competitive. The presumption, without justification, is that public enterprise is less efficient than the rest of private industry which it will come to match after privatization through the passage of time (p. 28). Consequently, on all criteria of performance other than temporary monopoly profits until competition comes along, no regulation is judged to be the preferred policy. In the meantime, regulation, over and beyond security of supply as guaranteed by legislation, is restricted to loose price control (p. 7):

> Competition is indisputably the most effective means – perhaps ultimately the *only* effective means of protecting consumers against monopoly power. Regulation is essentially a means of preventing the worst excesses of monopoly; it is not a substitute for competition. It is a means of 'holding the fort' until competition arrives.

This gives the industry the ability to develop more or less as it chooses in pursuit of profitability and represents no more than a token recognition of telecommunications as an essential service.

There are two interesting aspects of Littlechild's considerations which expose the true apologetic nature of his contribution. The first is that in making recommendations for the regulation of British Telecom, next to no analytical nor empirical material is presented which bears specifically on the telecommunications industry. Neither reader nor writer need know anything about the sector, except that it is being considered for privatization. Littlechild's conclusions depend on the abstract reasoning of the Austrian school alone, with its facile faith in the innovative and organizational benefits of management in private hands under the regulation of the *laissez faire* market.

Second, however, this purely abstract reasoning is abandoned in practice, in the short term, because of the static monopoly power that could be exercised by a privatized British Telecom. This pragmatism is hardly surprising, for *laissez faire* is not in general the preferred policy of either government or private capital. What is involved is an intense competition between capitalists over the telecommunications sector. An unregulated British Telecom would be unacceptable both to its potential rivals and to its business (the major) consumers. In the latter case, this includes regulation of the level of charges but much more is involved in the competition between capitals to gain access to a place in producing for the rapidly expanding, broadening, and changing telecommunications sector.

For Littlechild, the general faith in the workings of the market are raised to the highest level in a highly favourable assessment of the Stock Exchange as an agent of competition. If any enterprise is making excessive profits or incurring excessive costs (for example, through loose regulation), it will be taken over so that the process of acquisition operates as a mode of competitive control. This is so even where there is no competition in the product market, as Littlechild (1986: 5) argues for the privatization of the water authorities, where he recommends a mixture of discipline through a regulatory body, which can compare one authority with another, and discipline through the threat of takeover bids.

In the case of Littlechild, then, is found an economist who has shot to prominence as an economic adviser. There is a close affinity between his academic contribution and the politial ideology of the politicians that he serves. This is because of his fundamental commitment to the market and private capital even at the expense, one suspects, of a political democracy which is to be replaced by economic power mediated through the market place. This has its counterpart in the dismantling of the sources of Labour voters, as discussed earlier as an apparent element in Thatcherite political strategy. In place of public enterprise and other regulatory safeguards, public interests are best represented by self-funding charity and innovative monopoly. The negative results of monopoly power, which are potentially antagonistic to other capitals, are recognized but only at a secondary and temporary level. Specific recommendations for industries such as telecommunications and water are made on the basis of such abstract principles with a minimum of empirical analysis, and this extends to supporting propositions concerning the efficacy of the Stock Exchange and the presumed efficiency of British managers, once they are free to pursue profitability. Most fundamentally, there is a limited reliance upon analytical and empirical argument, whose rigour is suspect even where it is to be found.

The new synthesis

Whilst, in the light of the UK privatization programme, the neo-Austrian school, especially in the contributions of Littlechild, has resurrected old dogmas with limited innovative or supportive argument, the economics of public enterprise has blossomed more successfully in other directions. It has been given a new lease of life by privatization and this is confirmed by two telling examples. In the case of Rees' (1984) previously standard textbook, there are only two cursory references to privatization. Second, in the edited collection of policy contributions emanating from the Clare Group and brought up to date for publication in 1983 (Matthews and Sargent 1983), there is no mention of privatization.[9] Yet in Kay and Thompson's discussion of industrial policy in the collection on the *Performance of the British Ecomomy*, edited by Dornbusch and Layard (1987), there is mention of little else, other than competition policy and that is often more of the same. It might even be argued that industrial policy is being progressively reduced to competition policy, although Hay (1987) argues for a distinction between the two and for a coherent overall co-ordination across them.

The revival of interest in the economics of nationalized industries has been not only rapid but also prolific and all the more remarkable for the apparent absence of any substantial debate.[10] A synthesis (the term used here to describe it), paradigm, compromise or mode of rhetoric with the existing literature has emerged and a standard theoretical and empirical approach has been adopted. It is possible that the degree both of synthesis and of revival will be exaggerated since they are the responsibility of a relatively small number of authors, often working in harness, sometimes repeating the same thing in one or more separate publications. Most prominent has been the group in and around the Institute of Fiscal Studies and, judging by acknowledgements and contents in the contributions of other authors, they have been most influential. The other main group has formed around those who have contributed regularly to the *Oxford Review of Economic Policy* – see especially Vickers and Yarrow (1988) – although there have been regular partnerships in authorship between the two groups.

The purpose of this section is to give a point-by-point critical assessment of the synthesis. First, the explanation and justification of state economic intervention is based on the presence of market failure. This notion has been substantially developed at the microeconomic level beyond the original rationales for public enterprise – as being dependent on the problems of natural monopoly and externalities. Market failure today is now understood within a wide range of separately evolving areas of research – the theory of property rights, informational asymmetry and the incentive problems associated with contracts between principals and agents, the impact of positive transactions costs, the role of hierarchies, and other aspects of the new industrial economics arising from game theory applications and the theory of contestable markets.[11]

From this it must be presumed that market failure is endemic but there is no presumption that its existence warrants state economic intervention. This depends on the extent of market failure, its implications, and whether the (interventionist) cure is or is not worse than the disease. See Helm, Kay, and Thompson (1988: 42). In the latter instance, it is observed that the emphasis on market failure has been too one-sided and the mirror image of regulatory failure has to be considered.[12] Nor is the case for public ownership justified by substantial market failure, for nationalization is just one form of regulation or, more exactly, an institutional framework within which regulation has yet to be specified. More important is the extent to which the industry is served by competition and, where it cannot or should not be, how it is to be regulated. The associated attempt to achieve through regulation what proves impossible through the market is a central theme of the synthesis.

There seems little doubt that this broad approach takes as its organizing principle the notion that more competition tends to be beneficial, and underlying this is some general or partial equilibrium analysis with an associated Pareto equilibrium. This is the normal mode of operation of the economy from which there are greater or lesser divergencies. One particular result is required, that competition through more or potentially more firms is beneficial – what Weeks (1981) has called the quantity theory of competition. Such an outlook might be justified by the theory of the core, although the situation usually under consideration is one where the number of firms falls far short of that required to collapse the economy to competitive equilibrium, even if natural monopoly elements allowed this to exist.

Even leaving these considerations aside, the relevant move to oligopoly theory leaves matters notoriously indeterminate. Despite a proliferation of theories and approaches in recent years, encompassing behavioural motivations of managers (and unions), it would perhaps be extremely hard to find an economist who would be willing to swear that economic theory has shown that more firms lead to greater competition and improved allocative and productive efficiency. As Schmalansee (1988: 677) concludes in a survey article generally sympathetic to the new industrial economics:

> Recent theoretical research suggests that market conduct depends in complex ways on a host of factors, and the empirical literature offers few simple, robust structural relations on which general policies can be confidently based. Moreover, formal models of imperfect competition rarely generate unambiguous conclusions. In such models, feasible policy options usually involve movements *towards* but not *to* perfect competition, so that welfare analysis involves second-best

comparisons among distorted equilibria. In particular, there is no guarantee that making markets 'more competitive' will generally enhance welfare, particularly if non-price rivalry is intensified.

Significantly, then, in presentational terms, the synthesis has neglected mathematical analysis which is the usual mark of rigorously derived results by the norms of the modern economics orthodoxy. Relying upon informal notions of the beneficial effects of competition, it has no models and this cannot simply be because of their intractability, although this is a mighty constraint. For such models do exist as in Hagen's (1979) treatment of the second best for the public sector. One can hardly help but conclude that the neglect of the rigour of formal models derives as much from their agnostic or even unfavourable results as from their inherent difficulties.

Second, whilst formal models are eschewed for more intuitive argument, something else is borrowed from the new mathematical approaches to industrial economics. This is to cut away other forms of intuitive argument, those associated with long-standing traditions within industrial economics, emphasizing the role of institutional arrangements, their evolution, and the significance of behavioural factors. Moreover, a glance at a recent text of industrial economics, such as Clarke and McGuinness (1987), reveals a much richer series of considerations than those encompassed by the synthesis, and these are relevant to the issue of the impact of change in ownership as between the public and private sector – being concerned with the question, 'why firms exist and what influences their structure and development' (p. 166).

Third, just as market failure is the analytical starting point for considering state economic intervention so its incidence as monopoly assumes a central role. A basic conceptual distinction is made between natural and artificial monopoly, the former arising out of technical conditions of production, distribution and/or consumption, and the latter out of conditions of competition, such as entry. It is argued that the extent of natural monopoly can be exaggerated.[13] Its limits should be well defined and public or private provision appropriately regulated. For the rest, the economy is best served by modes of regulation that encourage competition where this, for whatever reason, does not arise 'naturally'.

In this, the theory of contestability has possibly played quite an important motivating role. This theory argues that even with natural monopolies, efficient allocation and absence of monopoly pricing are compatible with unregulated market provision, provided there is potential competition in the sense of rapid entry of other firms to undercut any excess profitability (Baumol 1982 and Baumol *et al*. 1982). In general, contestability has received quite a cool reception on the grounds that its prerequisite, the absence of sunk costs (and hence costless exit and the impossibility of effective 'incumbents') is far from realistic (Yarrow

1986: 342), for example.[14] Nonetheless, contestability provides a polar case in which the existence of a natural monopoly does not necessarily lead to the presumption of market failure, just as the existence of market failure does not necessarily lead to the presumption of the superiority of regulation. It is not so much the content of these results as their timing that is significant. They seem to reflect a shift in the orthodoxy away from a presumption in favour of public ownership and towards (regulation and) private ownership, very much in line with policy initiatives.

The very usefulness of the distinction between natural and artificial monopoly is, however, questionable as can be shown by use of the simplest partial equilibrium.[15] Natural monopoly would generally involve increasing returns to scale. This is most easily captured by considering overhead costs K and, otherwise, constant unit (and marginal) costs of production which will be taken to be w the wage rate. Accordingly, average cost of production for output x is $w + K/x$.

As only variations in wage costs are considered, these act as a proxy for all the other instances in which costs of production may be cheapened by confining a sector's production to a single firm. Suppose now that there is an alternative (small-scale) method of production which has no overheads and constant unit costs kw. Clearly, if $k \leqslant 1$, then small scale producers will always be most efficient. So assume $k > 1$.

Now natural monopoly is more efficient at high levels of output and, for efficiency, needs to charge a price w and receive subsidy K. Competition is desirable at low levels of output to prevent emergence of artificial monopoly, which would restrict output and charge price above kw. The level of output at which the two switch over is given by

$$w + K/x = kw$$
$$x = K/w(k-1)$$

This is illustrated in Figure 8.1. The important point is that the division between natural and artificial monopoly depends on the wage rate within the sector. It is not simply a 'natural' implication of technology and tastes (unless wages are taken to be an externally given parameter).

This is hardly a startling result, although it seems to have escaped attention in most discussion. Its significance lies in that a reduction in the level of wages implies a higher output at which natural monopoly first becomes preferable over a privatized industry. In other words, deterioration in workers' conditions improves the prospects of denationalization. Outside the world of perfect competition, in which wage rates are efficiently determined and from which the division between natural and artificial monopoly follows in equilibrium, the very distinction between the two becomes questionable as a starting point for analysis.

Figure 8.1 Natural monopoly and the wage level

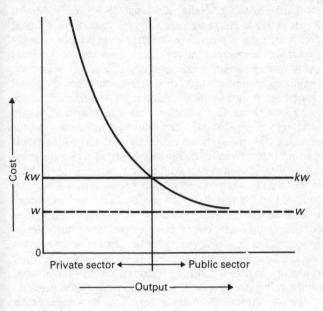

It has also been shown that a policy of privatization may not only be motivated by the wish to reduce wages but that this result may, *ex post facto*, justify privatization. Whilst empirical application of such a simple model should be undertaken with considerable caution, it can be observed that the nationalization of coal after the Second World War was at least in part motivated by the wish to unburden miners from the downward pressures on their wages and conditions that followed from the high costs associated with the fragmented structure of the industry in private hands.

The model also has an elementary extension to the problem of splitting the natural from the artificial elements of monopoly in the context of joint production and potentially inefficient cross-subsidization. Suppose to produce x and y jointly the cost is either $w(x + y) + K$ where K is the joint overhead cost or $kw(x + y)$ where now x and y can each be produced separately at unit cost kw. Clearly, whether there is a natural monopoly in the two activities together is given as before by whether $x + y \geqslant K/w(k-1)$. Once again this depends on the level of wages. Accordingly, downward pressure on wages and conditions can itself rationalize measures for hiving off certain activities previously undertaken jointly. It cannot be presumed that some of these activities are natural monopolies and others are not.

Fourth, the conclusion is drawn by the synthesis that the emphasis on state or private ownership of indusry, as suggested by a debate over privatization, is misplaced (see Helm, Kay, and Thompson 1988: 48, Kay and Silbertson 1984: 202, and Yarrow 1986: 324). Ownership as such is seen as a secondary issue. Of much greater importance are the conditions governing competition and regulation. Here, of course, there is a striking if unwitting parallel with the theory of perfect competition and general equilibrium in which regulation is redundant. For then also, ownership becomes irrelevant in the efficient allocation of scarce resources to competing ends. Rather factor inputs are, as it were, pooled and derive a scarcity price which is then delivered to passive owners as a revenue. In this way, the synthesis can be seen to be a generalization of, not a departure from, general equilibrium theory in which the latter is supplemented by regulation to deal with market failure. Nonetheless, the result of the irrelevance of ownership is carried through.[16]

This result is very much a product of the neglect of the issues of power and conflict. At times these are recognized, in discussing whether trade unions, for example, do better out of being employed by the state or by the private sector. The answer is ambiguous because at times the state's power can be used as a disciplining force – it has perhaps greater resources with which to hold out against industrial action – but at other times, it is more politically expedient to bend to trade union pressure or to allow the workforce to share in the benefits of unnatural monopoly.[17]

Such ambiguity, however, in the effects and, one might add, in the nature of ownership is not the same thing as its irrelevance or relative unimportance. Interestingly Kay (1987: 344) sheds light on this question by analogy with his own home ownership for which he is the asset holder, as recipient of residual financial claims on its value (even though purchase is bank financed), as well as the manager since all non-contractual use is subject to his discretion. But he does not observe that should the house be transferred to public ownership the nature of rights and obligations upon the owner (and manager as agent) become entirely different. More generally, the nature of ownership and the obligations and rights associated with it are different as between the public and private sector; some of these differences are formal as in the extent of Parliamentary accountability, for example, and others are informal and contingent as in the example given above of the balance of power in industrial relations. As Veljanovski (1987: 77) puts it within a property rights framework:

> privatization involves much more than the simple transfer of ownership. It involves the transfer and redefinition of a complex bundle of property rights which creates a whole new penalty–reward

system which will alter the incentives in the firm and ultimately its performance.

This also raises the behaviour and motives of those engaged in the industries more generally, not least the managers. Public ownership can instil an ethic of public service and this no doubt explains the previously low levels of remuneration of the managers of nationalized industries as compared to the private sector. That salaries have been immediately increased significantly for jobs and personnel, that have often remained much the same, would appear to be a small example of the erosion of an underlying commitment to public service.[18]

To put it formally, the synthesis reduces ownership effects to the changing constraints under which managers (and occasionally workers) optimize their utility, whereas the privatization programme is clearly intended to change the underlying 'utility functions' by moving them away from notions of public service and towards the exclusive goal of profit maximization. In contrast to this, Sen (1987b) has argued that the disciplines of economics and ethics should be reunited, not only so that economists can render normative advice but also in recognition of the economic content of ethical motives.

The poverty of the synthesis' understanding of ownership can be illustrated in another way – in its neglect of the theoretical and empirical literature which deals with managerial and worker motivation. This is not surprising because this literature tends to reside in the institutional tradition which the synthesis has rejected. Discussion of this is beyond the scope of this book, but it is significant that such problems have been addressed in the equally novel but less fashionable developments around Employee Share Ownership Schemes and Management Buy-outs. Here evidence is found for the significance of ownership in modifying motivation and behaviour, although participation in decision-making appears to be more important than participation in profits.[19]

The differences in the significance of ownership as between the public and private sector are non-trivial and this follows from the very different ways in which the private (even corporate) agent and the state relate to the economy. The putting aside of these differences by the synthesis follows from a fifth characteristic common to most orthodox economics – the procedure of separating out the economy from its social setting.[20] This does not mean that political processes, for example, do not enter at all, but that they only do so as external factors or as constraints on the economy, as in redistributional measures. But, as has been seen above, policies of privatization might have a very close connection to wages policy and may not be simply a natural consequence of production conditions. As Veljanowski (1987: xiii) observes:

> Having read a great deal of this literature I have reached the inescapable conclusion that, with some notable exceptions, very little has risen above the commonplace. One defect in much of the work stands out and requires comment. Privatization is a complex process which takes place in the commercial market place and in the political market place . . . one must examine the political and institutional context on an equal footing to the purely economic aspects.

Sixth, as a special case of the problems of a narrow understanding of the role of ownership, the synthesis sets aside the redistributional implications of privatization, even though wider share ownership has been one of its stated objectives. Quite clearly, however, changes in ownership have an effect on the demands for goods and on the distribution of welfare across the economy. Whilst such distributional considerations have been neglected in most areas of mainstream economics, this is justified in the synthesis by reference to the advisability of dealing with equity considerations through income support rather than through subsidy or direct provision.[21] Thus, it is better to provide the poor, for example, with income supplement rather than with cross-subsidized services. For otherwise this will lead to allocative inefficiency as the poor could be made even better off, at less expense to the rich, by their spending their subsidies as they choose at prices which reflect marginal costs.

Whilst it might be accepted on the narrowest of technical economic grounds that the nationalized industries, through providing cheaper services, are not necessarily the most efficient means of pursuing social objectives, this does not guarantee that redistribution will indeed occur through other mechanisms such as income support. As has been argued in the case of the Kaldor–Scitovsky compensation criteria, to judge one state of the world to be superior to another, because the poor could be more than compensated for their inferior position within it, is of little comfort to the poor unless compensation is actually paid.

Thus, Dilnot and Helm (1987) do discuss the role of energy as a merit good, but in terms which are predominantly distributional and deserving of targeting to the poor. They recognize that certain absolute standards of provision may be worthy of guarantee by the state in line with their interpretation of Sen's (1983) notion of capability which addresses the ability to participate fully in society – food and shelter being a basic minimum. In doing so, capability does tend to assume the role of a minimum standard of individual consumption. This has two effects. One is that the social effects of privatization are reduced to the issue of distribution and this is further confined to a focus upon the poor (as if the effect on the bulk of the population, as well as the minority of the rich and powerful who become more so, were of no significance).

The other effect is to set aside the issue of capability as a social property, a matter on which Sen appears to remain ambiguous. Once, for example, society moves beyond the minimum standards of survival, individual capabilities are profoundly social – the right to join a trade union, for example. Privatization cuts across and affects these social capabilities (as well as the distribution of individual consumption and income). These factors are necessarily neglected by Kay and Vickers (1988: 302) for whom:

> Industrial policies generally, and regulation in particular, are usually ill-suited to wider distributional ends, which are better accomplished by other instruments of public policy.

In contrast to the dictum quoted by Sen (1987a), following Brecht, 'grub first, then ethics', the synthesis would have, 'first economics, then ethics, then grub'! It is a recipe for poor economics, limited ethics and deprivation.

Thus, nationalized industries with subsidized prices for necessities or to certain sections of the community may be a much more certain form of provision than compensating income following privatization and/or the withdrawal of subsidy. It is perhaps no accident that the current policies of privatization are accompanied by those in the arena of welfare that are, if anything, compensating in the opposite direction, not least the reform of social security payments and the abolition of the higher rates of income tax. As such, consideration of the issue of redistribution tends to reinforce the earlier criticisms of the synthesis in that it oversimplifies the relation between politics and economics, it neglects the issue of power and conflict, and it reduces the question of ownership to the rights to a stream of income.[22]

In addition, the simple model above can illustrate that the relationship between redistribution and privatization is not so simple. Suppose, for example, that government seeks to lower subsidized output in a public sector industry. The implication is that output may possibly fall below the level of viability of the natural monopoly, thereby rationalizing a decision to privatize. This may seem far-fetched in the context of the simple model, but consider the fate of British Leyland, for example, whose level of output has now fallen so low that its future as a mass producer of cars is doubtful, certainly in the absence of joint ventures with other companies, although this is in part the result of changing technical imperatives dictated by scale economies.

Seventh, the synthesis' analytical principles are coupled to a particular view of the history of, and prospects for, the nationalized industries. Whilst the difficulties of measuring the performance of nationalized industries are recognized, the general conclusion is drawn that this is inferior to the private sector in the presence of competition between private and public enterprise. Similar conclusions are drawn for the, admittedly as yet short-term, consequences of those privatizations that have been implemented.[23]

On the other hand, it is recognized that the theoretical analysis only provides limited guidance to the study of a particular industry because of the complexity of the considerations to be taken into account. Each of the forms of market imperfections is potentially present, the industry has an historically evolved structure which affects future prospects through, for example, the presence of 'incumbents' who may impose entry barriers,[24] and the vertical as well as the horizontal structure of industry may be of importance, as in control over the different areas of production, distribution and sale – as in energy, for example.[25]

By implication, and often explicitly, the synthesis presumes that nationalization occurs historically as a result of the incidence of natural monopoly.[26] In fact, the post-war nationalizations were very much the product of a fragmented industrial structure, unable or unwilling to be reorganized through the market under private hands. In so far as natural monopolies do exist today, as in energy distribution networks, these are the positive effect to a large extent of public sector management, whatever other deficiencies may have resulted. Similar considerations apply to the previous role played by the public corporations and the damage likely to be done under privatization to the provision of collectively beneficial services such as training of skilled workers and the pursuit of R&D, although the public sector record has not been entirely satisfactory.

Nor has the experience of the nationalized industries been one of slow realization that the optimality conditions as suggested by welfare economics were not being implemented (Helm, Kay, and Thompson 1988: 49, Molyneux and Thompson 1987: 50). Rather, whilst this may be one way of reading successive White Papers on the industries, it exaggeratedly honours the role of economists. They have been slow to study how the public sector has operated, focusing instead on how they think it should operate in an ideal world. Accordingly, the paths of the public sector and of public sector economics have always diverged (Fine and O'Donnell 1985), and it is only relatively recently that the gap between these paths has been narrowed by closer considerations of what the industries' objectives should be and how they should be achieved in the absence of the profit motive and the threat of bankruptcy and takeover.

In assessing performance, some more or less sophisticated attempts are necessarily made to measure the movement of total factor productivity over time and to compare this with the performance of the private sector. Many of the attendant difficulties and reservations in doing this are remarked upon (Molyneux and Thompson 1987: 53). Byatt (1985) also reasonably suggests that assessment needs to be made over an economic cycle, for private capital might be expected to be keener on public service provision during recession when private profit-making opportunities are in shorter supply. How long a cycle should be taken is, however, another matter, particularly when the longer-term health of the economy is under consideration. Even so, it is far from clear whether short-term exercises have the desired effect of isolating and measuring the impact of public as against private ownership. For the former presumably have had social obligations that have acted as constraints and which have been absent from the private sector. An appropriate exercise might be to assess how the private sector has or would have faired under the terms of operation of the public sector!

Also the differences across industries and firms will be the consequence of many factors other than ownership alone and this applies to comparisons within public and private sector industry as much as between them. An appropriate assessment might be made of whether relative performance within the private sector shows more or less divergence than performance between it and public sector.

The measurement of changed performance over time and the assignation of differences between the public and private sector to differences in ownership implies that entrepreneurship is being tested by these means. This certainly ought to include an assessment of the ability to innovate in the widest sense, as has been accepted by many for the major debate over British entrepreneurial performance around the turn of the century.[27] Yet the theory that the synthesis brings to bear on the relative performance of the public and private sectors is almost entirely devoted to the matter of static allocative and productive efficiencies. It has not been concerned with the dynamics of entrepreneurial performance which must surely be a major factor in explaining differences in the growth of total factor productivity. Crafts (1988), a noted economic historian of Britain's economic performance, doubts whether privatized industries will invest sufficiently, and low levels of investment are universally recognized to be part of the British economic 'disease'.

Indeed, the only way theoretically in which the synthesis tends to allow past performance to affect the future is through the idea of 'incumbents', which Crafts also considers will be too influential. Those with capital sunk in the industry may have a competitive advantage that will have to be eroded through easing entry before giving way to less strict regulation. This suggests that any static and stable equilibrium, to which the industry may be attached, is liable to be dependent upon the initial ownership structure. But, equally important, it follows that the industry's position is dependent on the continuing dynamics of entrepreneurship which both establishes future incumbents and the (shifting) position they occupy. Yet the synthesis is preoccupied almost exclusively with the deadweight static implications of regulation and competitiveness.

Eighth, and emerging in the previous paragraphs, the synthesis has a central preoccupation with static and predominantly allocative and efficiency concerns. As Ulph comments on Kay and Vickers (1988: 348):

> The whole presumption which underpins the Kay/Vickers paper, and much else in this area – that the more we can make the market look like a perfectly competitive one the better – may simply not be true in a dynamic context.

It simply needs to be repeated that it may not be true in a static context either.

Ninth, this focus on the static leads to a rejection of other theories, some of those within the more descriptive tradition of industrial economics having already been noted. Engaging with others, with the occasional exception of the neo-Austrian school, is extremely rare, on the grounds presumably of targeting those that are currently most influential on government policy. The neo-Austrian's view that the market works best to encourage enterprise and the dynamic benefits of change is seen to be too one-sided when set against the potential static inefficiencies of monopoly (Helm, Kay, and Thompson 1988: 51). Thus, when Beesley and Littlechild (1983) argue that regulation will only be needed in the short run after privatization, until such time as the monopoly entrenched under state ownership has been eroded, this is dismissed (Yarrow 1986: 343).

More subtle has been the erosion of some of the traditional concerns of public sector analysis. Rees (1984), for example, puts forward four objectives for nationalized industries. Two are to correct market failure and to redistribute, which the synthesis recognizes, if only to leave the latter to other agencies. But the other two, of centralized long-term economic planning and to change the nature of the economy from capitalist to socialist, are entirely ignored. Accordingly, the synthesis is confined to a narrow microeconomic conception of what constitutes the role of the nationalized industries.[28]

This has its counterpart in the setting aside of a whole series of areas of importance, reflecting the narrowing of the scope in which the analysis of the nationalized industries is situated. Consider macroeconomic policy, for example. Levels of investment, inflation, balance of payments, and unemployment are all potentially influenced by the plans of the nationalized industries. Whilst it might be argued that such objectives better belong to the world of more efficient instruments (that do not distort the optimality conditions at the microeconomic level), practical knowledge of such policies remains elusive and, as has been seen in the discussion of redistribution, cannot be relied upon to be implemented in an imperfect world. And the world is extremely imperfect, given the persistence of macroeconomic disequilibrium. Yet, to put it bluntly, it is as if mass unemployment, for example, does not exist for the purposes of the synthesis, a feature, unlike its response to policies of privatization, which makes it unresponsive to contemporary economic conditions.

Similarly, the narrow focus of the synthesis tends to preclude policy towards a co-ordinated expansion of those public sector industries that are intimately linked through the input–output table. As Beynon *et al.* (1986) have shown, the fate of the coal, steel, and water industries are closely tied together (and the same must apply to regional policy through these industries and their associated multipliers).[29] In contrast, the synthesis pushes towards increasingly microeconomic decision-making where this can be appropriately corrected for market failure through regulation. It must be questioned whether the net of failure has been cast wide enough.[30]

This is not necessarily to go the whole hog and to demand a fully, centrally planned economy. But the neglect of the relationship between nationalization and socialism (and the associated conflict between capital and labour both at economic and political and at micro and macro levels) is symptomatic of many of the points that have been made. Nor is this commentary to be dismissed as an ideological gloss over political motives or woolly thinking. For, whilst public ownership has meant many different things to different commentators, from the weakest commitment to the mixed economy through to full central planning, this has been because of economic analysis and objectives that differ from those of the synthesis, in many instances along the lines that have been indicated. These considerations cannot reasonably be excluded if an appropriate history and future policies for the nationalized industries are to be constructed.

Finally, and tenth, there is a policy synthesis. Neither privatization nor public ownership are opposed for the form of ownership is less important than the level of competition and the nature of regulation. On the other hand, the privatizations that have been implemented are criticized for having introduced negligible levels of competition, as in the selling-off of British Gas as a single company and the limited scope given to Mercury, let alone others, in telecommunications. Policy in the synthesis' view should be formed through a three-stage analysis. Examine the industry for presence of natural and artificial monopoly. As these are mixed in practice, target them, respectively, by a balance of regulation and inducement to competition (Helm, Kay, and Thompson 1988: 47). The last stage is to ensure effective monitoring of performance through an appropriate system for the feedback of information to regulators.

The synthesis is understandably peeved at being totally ignored in the implementation of the Government's privatization programme. It is reasonably recognized that motives other than the introduction of greater competitiveness have been decisive, such as the higher selling price for a privatized and liberally regulated monopoly. On the other hand, the synthesis has allowed the Government to set the agenda for debate. For it has confined itself to an analysis of those sectors which have been targeted for privatization. The logic of its analysis, however, points to a comprehensive review of the economy as a whole, sector by sector, including those sectors in which the state does not have a substantial stake, to see whether an extension or modification of regulation or even public ownership is required. In the absence of this, the synthesis will continue to be little more than a self-appointed watch-dog and champion for competitiveness where the Government chooses to intervene.

Tailing upon the Government's plans for privatization indicates a deeper problem with the policy proposals of the synthesis – it has no broader perspective on the British economy in which to situate its suggestions for the public sector. This follows from the limited view taken of the role of the nationalized industries in the context of macroeconomic planning and in terms of industrial planning across sectors. It is complemented by an apparent neglect of the specific longer-term problems of the British economy. The relevance of the comparative performance of the public and private sectors has to be questioned when it is the poor absolute performance of each that stands out in international comparisons. Until an explanation is offered for the relative decline of the British economy, differences in policy over the degree and nature of regulation of the public and private sector must be considered to be of secondary importance.

Concluding remarks

To summarize: the synthesis is essentially concerned with static efficiency through a division between natural and artificial monopoly targeting each, respectively, by regulation and market forces. Privatization is understood primarily within a narrow economic framework, thereby diluting the significance of ownership, and policy is based on the notion that competition is favourable where possible.

With major privatizations in place, a second wave of literature has now emerged around the issue of regulation.[31] Analysis of this provides, in many ways, the end result of the synthesis' approach, and so it is examined here by way of conclusion to this chapter. The subject is by no means new, as it has previously been developed in the United States in response to the regulation of private utilities, especially electricity.[32] The problem posed is how to devise the appropriate number, choice, and level of target variables so that a regulator can guide and monitor companies through a corresponding system of penalties and rewards.

The commanding heights

In this, it has surprisingly gone unnoticed that there is a comparable problem in the literature concerning market socialism and the direction of a centrally planned economy. There is also an affinity with the literature on indicative planning. Significantly, however, these contributions tend to approach the problem from the opposite point of view – how to achieve previously centrally determined objectives through decentralized management rather than how to bend privatized concerns towards an unknown efficiency objective. Perhaps it is unsurprising that the separate literatures have yet to confront each other.

Here, it is not possible to review the regulation literature for it is both extensive and far from coherent. The problem is generally posed against a background of asymmetric information and objectives as between government and managers (and often workers, consumers, taxpayers and any other interest group considered relevant). How is efficiency, as against monopoly, to be achieved without dulling incentives? Focus is usually on the use of indices such as $RPI - X + Y$, where X is a price decrease to reflect productivity increase and Y is a price increase to accommodate abnormal movements in the prices of essential inputs. By analogy with the planning literature, it follows that this is liable to lead to distortions. In the case of British Telecom, for example, an overall price increase allows for differential increases across the various categories of calls to extract maximum revenue. More generally, as is well known, to achieve undistorted targeting on a specified number of variables, an equal number of instruments is required.

At an intuitive level for the well-trained economist, the primary focus on regulating prices is puzzling. As there is a duality between price and quantity (as between cost and production functions), so there ought to be a corresponding duality between price and quantity regulation. As it were, regulating price must involve an implicit assessment of the conditions of supply and demand, for which there is a corresponding rule for regulating quantity in the circumstances specified.[33] Another example might be given by the duality between specifying a target rate of return and a target level of investment (in the simple case, for illustrative purposes, when there is an inverse relationship between the two).

The greater favour shown to price regulation follows from an asymmetry in the duality between quantity and price in all but abstract theory. Once outside the static world of neoclassical economics, the setting of quantities is prior to their sale. As is intuitively clear, this makes quantity regulation more interventionist than price regulation. Would British Telecom, for example, prefer to have an externally given price or an externally given quantity of calls that must be achieved through the price set?

Discussion has not, however, been confined to price regulation alone.

Kay and Vickers (1988) distinguish between structural and conduct regulation. The former governs conditions of entry and, in the context of privatization, the initial industrial structure *ex post facto*. For major denationalizations, such structural regulation will be imperative and cannot be confined to the settling of who will be the incumbents on the starting grid. Structure will have to be reviewed from time to time in view of technical and other developments and as the industry reorganizes itself through acquisition, merger, and diversification.

Conduct regulation, most obviously in setting prices but also embracing other rules and requirements such as obligation to supply, suffers from the previously mentioned surfeit of targets relative to instruments. This can lead to polar opposite outcomes. In case of investment, a guaranteed rate of return encourages over-investment and excessive capital-intensity as such costs are subject to sure mark-up. On the other hand, if the regulator attempts to optimize by setting price equal to short-run marginal cost, this will induce under-investment since firms will be penalized for unanticipated excess capacity by being forced to lower prices and possibly failing to cover overheads.

This all suggests an evolving relation between firms and regulators over time – in formal terms this is accommodated by models of repeated games between the two. In practice, there is the danger of regulatory capture, in which the regulator loses independence and becomes a mouthpiece for the industry. Who, to paraphrase Karl Marx, is to regulate the regulator?

These comments all concern static considerations in which the regulator attempts to set appropriate targets and to obtain relevant information for doing so. The presumption is that the firms are better informed and have an incentive to report falsely. Schemes such as yardstick competition have been devised to deal with this, although enhanced democracy and representation of workers and consumers have generally been neglected. However, when all of these problems are translated into a dynamic context – how to innovate and generate productivity improvement, for example – there is a notable lack of discussion and an implicit acceptance of the neo-Austrian proposition that this is best left to entrepreneurial initiative, with an imperfect regulation to moderate excessive profits.[34] It must be recalled that this last condition would be rejected by the neo-Austrians on the grounds that entrepreneurial energy would be blunted if excess rewards were regulated away.

The conclusion to be drawn from this discussion is that a theory of regulation depends upon setting aside major problems of industrial development or, more exactly, leaving them to the market, focusing instead on the limited scope for a regulator to enhance efficiency in face of informational deprivation.[35] In effect the regulator becomes an impoverished instrument of industrial policy and only by narrowing down what this encompasses can regulation be perceived to be potentially successful. For, at the end of the day, the objective is little more than to obtain a degree of Pareto efficiency in a world falsely represented as a static partial equilibrium (or divergence from it).

If the problems listed above, and others, are brought together and placed in a dynamic context, then the regulator is repeatedly assessing industrial structure and reorganization, cost and price, and informational abuses and lacunae. Consequently, even to undertake regulation in a static context satisfactorily, the information if not the powers to direct the industry will be available. For it is questionable whether the information required to regulate an industry satisfactorily is so much less than that required to command it (or regulate it, as it were, through public ownership).

That is, unless there is an asserted belief in the absolute merits of private enterprise. Such is the basis for regulation, as is recognized by Carsberg (1986: 82), Director General of Oftel. For whilst he acknowledges that the imperative to privatize may have some roots in technical change, he does not address the issue of how this is to be sustained after privatization. Instead, all that is offered are two principles, insisting on the superiority of private over public enterprise:

> First, a government may be unable to resist political pressures to use its ownership rights to interfere with prices or it may move against procedures which are designed to prevent anti-competitive practices . . . . Secondly, and perhaps more importantly, reasons exist for believing that managers behave more entrepreneurially in the private sector than under state ownership.

Such beliefs, and their implied but unproven favourable effects on the economy, are ultimately the limited rationale both for privatization and for the reduction of industrial policy to the parameters of industrial regulation.

Chapter nine

Privatization and property rights: from electricity to coal

In chapter 8, a critical assessment has been made of the economic analysis that emerged in response to privatization.[1] The prominence of the new orthodoxy, termed the synthesis, contrasts sharply with the shallow analytical and empirical support to be found for the neo-Austrian school which, with more than a dash of pragmatism, rationalizes Government policy through compromise with its analytical principles. Whilst the synthesis is able to expose the limitations of Government thinking and policy making, it does so on too narrow a basis. For there is sore neglect of the historical and political aspects involved, as well as the broader economic considerations covering the weakness of the British economy and the structural causes of inadequate industrial restructuring. As Holmes (1988) put it:

> The myth which informs the privatisation of the Electricity Supply Industry – and indeed privatisation in general – is that the nationalised industries are somehow an accretion on the body of the nation, which can be destroyed and then reconstructed at a stroke. The truth is that the state industries, for better or worse, are a reflection of the way that this country works. They have evolved over decades, and are not going to be changed overnight, by privatisation or anything else.

This chapter begins with the privatization of electricity supply, the first section by demonstrating the continuing commitment to nuclear power, and the second by ranging across the deficiencies in the Government's privatization proposals. The conclusions to be drawn are that there is a policy continuity with the past, despite the break with public ownership; as in support for nuclear power; in so far as energy efficiency and conservation are neglected along with the factors conducive to poor performance in power station construction; research and development will lack coherent co-ordination and adequate support; and the elusive phantom of competition will be the grounds on which the merit order in generating electricity is liable to be sacrificed with potential costs of £1 billion yearly.[2]

Discussion of the privatization of electricity serves two purposes as far as the coal industry is concerned. First, the intimate relationship between the two sectors means that there are potentially profound implications for the coal industry following on from the privatization of electricity. The most obvious of these is the importing of coal. The proposed building of two new ports on the Humberside, for example, with the capacity to import 8 million tonnes of coal annually, threatens 15,000 jobs in the British coal industry. The broader implications of this in terms of damage to the balance of payments seems to have been set aside by official thinking.[3]

This is just one aspect of the macroeconomic implications of pit closures. More generally, especially within a Keynesian framework, it can be recognized that there are quantifiable effects of such closures which mean that the net consequence is harmful to the economy. This follows from what has unfortunately been termed the social cost of closure – the need to make redundancy and dole payments, the loss of direct and indirect taxes, etc. These social costs are all too readily mixed up with the less quantifiable, but no less tangible, effects associated with the decline of a coal-mining community.[4] However, unemployment and its associated costs are subject to calculation and should form the basis of policy making.

This, within the broader context of the ideology of uneconomic pits, is the subject of the third section. Discussion of the type to be found there has been presented as evidence at a number of open public inquiries, such as Sizewell, colliery closure reviews, and the Committee Stage of the Bills proposing the Humberside Ports. Appendix 9.1 reproduces the evidence as it was presented at the review of Cadeby Colliery. At no stage has the substance of such evidence, whether in principle or for particular pits, been seriously challenged.[5] Yet, in evidence to the Energy Select Committee (1985/86 ii: 7) when asked, 'Can your department be confident there have been no pit closures where the net benefit of gain by the National Coal Board has been exceeded by the wider social costs that have resulted?', the then Energy Minister Peter Walker replied, 'Yes, I would be very confident of that.'

These factors all point to the failure of the government to provide a satisfactory economic basis on which to justify the privatization of electricity. Nor is there an apparent empirical rationale, for as Fare *et al.* (1985) conclude after calculating six different efficiency measures, there is little difference between public and private sector performance in electricity supply with, if anything, the former having a slight advantage. This does not mean that there is no rationale behind privatization policies. One that appears to be prominent is the wish to run down the British coal industry in order to weaken the power of the miners and of the trade union movement more generally. This suggests that the policy

The coal question

to privatize the mines, announced in late 1988, may also be without a direct economic rationale, and this is the second lesson that can be learnt for coal from the privatization of electricity – its fate may well be decided by the same levers of political strategy or prejudice, depending upon your viewpoint.

The fourth section discusses the miners' strike of 1984/5, exploring themes around the issue of ownership that were previously aired in chapter 8. There, it was implied that ownership is a complex and shifting aspect of the economy, incorporating non-economic factors. This is borne out in the case of electricity. John Baker (1989), for example, future chief of the larger of the two proposed generating companies under privatization, has already questioned the collective principle of ownership in so far as it guaranteed supply:

> We need to define ways of running our power stations so that we can exploit our power contracts profitably. Our task will not be to keep the lights on whatever the cost. It will probably pay us to ensure we never stress our plant.

And much the same applies to the collective and individual objectives of those associated with the company:

> Again, notice how different the vocabulary is from that which we are used to in the CEGB . . . . The job isn't about shouldering national responsibilities but about meeting contracts, improving profitability, about seeking out opportunities but only exploiting them if it pays to do so . . . . All this requires us to change the way we think and change the way we behave and change the way we do things.

This change, change, and yet more change might not be thought to be so much of a dramatic shift in the nature and exercise of property rights. That all depends upon the conflicts to which it gives rise and how they are resolved. But when these do become less delicate, the question of rights is extremely prominent – as is illustrated by the miners' strike. If, in retrospect, privatization of the industry was at stake, it is hardly surprising, given the other issues involved, that the conflict should have been so keenly fought.

Finally, in the concluding remarks, a partial comparison is made between the strike of 1984/5 and the General Strike of 1926, more to bring out the points of difference than the obvious parallels. For some, the strikes are to be seen as disturbances in the otherwise smooth evolution of the coal industry and economic and social life more generally. For others, each proved a fundamental conflict and potential turning point in British society. In either case, each raised a range and complexity of issues far beyond their immediate focus of struggle.

The enigma of nuclear

In official attitude, there has been a dramatic turnaround in the assessment of nuclear power in the wake of the proposals to privatize electricity. After the Sizewell Public Inquiry which lasted from January 1983 to March 1985, Sir Frank Layfield, the Inspector, produced an eight-volume report some eighteen months later in December 1986 that essentially approved the building of a PWR nuclear power station both on economic and safety grounds. In the interim, there had been dramatic falls in the price of oil, with the price of coal following, and the Chernobyl disaster had cast further doubts on the safety record of the nuclear industry. The Inspector had not been allowed to take these factors into consideration as, of a necessity of their timing, there had been no evidence presented on them. Approval was given by the Government to the CEGB for the building of the Sizewell station. The basic economic grounds were those of the cheapness of nuclear as opposed to coal-fired generation.

In preparations for a speech to his managers John Baker (1989), the much-admired leading advocate for the CEGB at Sizewell who was to be rewarded by heading the National Power Company (incorporating all nuclear supply in the privatized electricity industry of England and Wales), had already taken an entirely different view:

> If we take away the costs of the Grid, and allocate to Power Gen their share of the CEGB's costs and resources, and then allocate all the remaining costs, including overheads and profit, fairly to our fossil stations and nuclear stations respectively, then broadly speaking we can see that in 1989/90, a fossil unit costs on average about 3.5p/kWh, of which 2.0p is fuel. A nuclear unit costs about 5.0p per kWh, of which again about 2.0p is the nuclear fuel cycle.

At last, representatives of the industry were admitting what had been increasingly argued over the years,[6] that nuclear power was uneconomic, not only in the past but also into the future. The proponents of the power supply that from the 1950s onwards was going to be too cheap to meter had at last come of age and to their senses.

But celebrations over this analytical triumph could only be limited. Policy was not to change even in the coming era of privatization. For the Government's intentions were to ensure that at least 20 per cent of supply would remain nuclear even though such power stations were commercially unattractive. In place of economics, diversity of fuel supply became the most prominent rationale for the commitment to the nuclear power programme. This is to be paid for by a nuclear levy on consumers, to guarantee that none of the excessive costs fall on the industry and potential investors, with the Government liable to back the building of four more PWR power stations prior to privatization.

This is the latest in a long series of analytical acrobatics whereby policy decisons are justified. In 1981, the Energy Select Committee (1980/81) had revealed that the CEGB's methods for supporting nuclear power on economic grounds were riddled with problems, not least the use of historic costs for future planning, the assumption of the most optimistic completion times for power station construction, the comparison of nuclear on full load with coal not so, etc. Such criticisms met with responses which merely served to strengthen the case presented by the CEGB without at all affecting its conclusions of preference for nuclear. Thus, at the Sizewell Inquiry, assumptions were made about the future trend of coal prices which were sufficiently high to favour nuclear over coal for electricity generation. Meanwhile, preparations were soon in hand to import coal at depressed international prices and to force down the price paid to British Coal. Effectively, the Department of Energy was using one (high) future price of coal to justify the nuclear power programme and another (low) price to run down the coal industry by pointing to its pessimistic prospects.

Revealing the source of the British commitment to nuclear power is beyond the scope of this book.[7] It is crucial to recognize how much it is out of line with developments around the world. As the *Energy Economist* (1988a) reported, Italy, Netherlands, Finland, Austria, and very probably Switzerland are dropping out of further reliance upon nuclear power, joining Norway, Denmark, Greece, Portugal, Eire, Luxembourg, Monaco, etc. The United States has not commissioned a new station since the early 1970s, before the accident at Three Mile Island which had the additional effect of encouraging cancellation of existing orders, even those ready to switch on. Only France and Belgium in the western world remain firmly committed to nuclear power with Spain, West Germany, and the UK on the borderline in western Europe. The reasons for this are not primarily due to safety considerations in the wake of Chernobyl.[8] The *Energy Economist* (1988b) reports a near quadrupling of operating costs in the decade from 1974, so that these now exceed the price of the coal equivalent, even though nuclear was presumed to compensate for its higher (and expanding) capital costs through lower fuel costs.

Thus, the preference for nuclear power embodied within the privatization programme has a limited economic rationale into the future just as much as the savings promised in the past have failed to materialize. But the economic damage imposed by a nuclear programme is not limited to these direct cost considerations alone. For there are also profound implications for R&D, also as in the past. This has to be seen against two crucial aspects of the British economy. First is the extent to which Government R&D expenditure has been disproportionately devoted to the sectors around defence and aerospace with limited application and spin-off to the economy more generally. The second is the dominance of nuclear power within the R&D programme for energy. As the Energy Select Committee (1987/88b) observe, the imbalance between resources devoted directly by Government to coal and nuclear power is astonishing, with £209 million being planned for the latter in 1988/9 compared to £1.4 million for the former. This is equivalent to 5p per tonne of coal burnt in the UK as compared to 31p in the USA, 34p in West Germany, and 105p in Japan.[9]

Clearly this is not a novel observation, but it has to be set against the background of privatization and the wider problems of the British economy. Obviously some reassessment of priorities is taking place with the decision to phase out the fast breeder reactor at Dounreay. But this is simply a cost-cutting exercise which has its counterpart in the coal industry (only at a much lower scale of expenditure) with the Grimethorpe pressurized fluidized bed combustion plant (clean coal-fired) placed in jeopardy for lack of UK government funding. Cutting expenditure is not to remedy the long-standing absence of an overall co-ordinated strategy either for energy or for R&D (except to leave it to the market place).

A lack of coherence is equally illustrated by recent developments around the UKAEA. Under financial pressure, it is seeking to diversify commercially into non-nuclear activities. But is this the only appropriate policy for the self-proclaimed largest R&D organization in (western) Europe? Surely it ought to be playing a central role in a coherent strategy for the development of new technology for the economy as a whole?

Privatization of electricity

At the time of writing, discussion of the privatization of electricity had gone through three stages. Initially, there was some discussion of whether denationalization was worthwhile or not.[10] Once the Government's intentions were made clear, this gave way to contributions around the new structure for the industry that could be adopted after privatization.[11] For the purposes of outlining proposals, the system of supply was divided down into the three stages of generation, transmission, and distribution, corresponding to the public sector system of the CEGB as both generator and as operator of the national grid and the twelve area boards responsible for sales. The different schema were then concerned with the ownership structure within and between these separated activities. Should ownership of the grid, for example, be by an independent company or be subject to the control, other than through the market, of those responsible for either sales or generation?

Following the Government's White Paper committing it to privatization of electricity,[12] this issue was resolved in practice by the preservation of the twelve area boards responsible for sales, their joint ownership of the grid, and the division of existing generation into two separate companies, to become known as National Power and Power Gen. Provision was to be made for independent companies and for the area boards to generate electricity and, where appropriate, for it to be contracted to or through the grid company to customers, possibly other than the area boards. Subsequently, the third wave of literature has been concerned with the most appropriate form of regulation of the industry. Particularly prominent, and distinctive for electricity, has been the issue of how to cover the costs of nuclear power, given the imposed requirement that this should essentially constitute 20 per cent of supply.[13]

The debate over privatization of electricity has been very much within the terms presented and critically assessed in chapter 8. As such, Government policy has been driven by a neo-Austrian approach tinged with pragmatism and opponents have been located within the outlook of the synthesis – where competition and regulation are perceived to be central. In this regard, the achievements of the synthesis may be considered to have peaked and even to have been surpassed by the Energy Select Committee's (1987/88a) response to the Government's privatization plans which, scarcely reading between the lines, can be seen to have concluded that the Government's plans are arbitrary in principle for having been convinced of the virtues of privatization without having worked out their effects in practice; arbitrary in practice in the different treatment of the ESI in England and Wales as opposed to Scotland; afford an undue preference to nuclear power especially over coal; and treat energy efficiency with scant regard. In addition, the Select Committee seeks extensive and strong regulation. Thus, in summary (p. xix):

> The White Papers raised as many questions as they answered and were especially thin on detail of such crucial areas as the economic justification for vertical integration in Scotland but not in England and Wales, or how the merit order can be replaced effectively . . . or what precise functions the Regulator would fulfil . . . many matters remain opaque . . . and we suspect . . . daily discovering new problems which need to be solved . . . . We cannot make recommendations about policy when policy is unknown . . . the Government . . . runs the risk of producing ill-considered, spatchcock legislation. Electricity is too important an industry for the country to gamble that everything will come out right.[14]

There is, however, an inevitable lack of realism about these strongly worded criticisms and demands and not just because of the limited account that the Government, however much embarrassed, might take of the Report as the Committee well knows from previous experience. In

Privatization and property rights

the case of regulation, the Committee can easily be interpreted as asking Ofelec (or other agencies such as the Monopolies and Mergers Commission) to formulate and implement both energy (and, by implication, industrial) policy to an extent and with a coherence that has previously been absent in Britain over the whole post-war period! The paradox is that this is being asked of a Government that poses as seeking to relieve itself of such responsibilities. It is hardly going to welcome such extensive duties being taken up by a 'quango' that it is itself responsible for appointing.

From the events and debate surrounding the privatization of electricity, the Government appears to have adopted the strategy of 'go forth and privatize', with a dogmatic belief in its virtues and the intention of resolving problems as they arise.[15] This, with some justification, is seen as a sharp break with economic policy of the past. It would be a mistake, however, not to recognize the extent to which privatization itself represents a continuity with the policies of the past, in terms of the increasing commercialization of public corporations' operations, especially in policy for energy and electricity. It is this that is emphasized here.

Consider, for example, the issues of energy conservation and energy. Hillman (1984: 43) reports that between 1979 and 1984 the government invested as little as £6 per household on insulation and £12 per household on energy efficiency. This is extremely low by international standards; France, West Germany, and Holland spent six to seven times as much per capita in 1982, and Denmark fourteen times as much. Even after an increase in staff of ninety at the Department of Energy, the French equivalent to the UK Energy Efficiency Office had 500 central staff and 2,000 in provincial branches. The French and West German governments give three times as much support to industry for energy conservation as the UK, and the average for the then ten member states of the EEC was still double that of the UK government.

This is the sort of situation that existed prior to privatization, one in which, as Hillman and Bollard (1985) describe it:[16]

> In marked contrast to a programme of increasing energy supply through power station construction, a programme of energy conservation has a short lead time, is labour intensive, does not require a major financial commitment over a long period but will provide long term benefits.

There can be little doubt that deficiencies in energy conservation and efficiency will be worsened after privatization. Despite the poor record under public ownership, incentives to conservation and efficiency depend upon suppliers benefiting more from the cheaper energy services that they provide than they lose in reduced demand (Kahn 1987). Given new entrants must provide capacity and sell it, this is not liable to occur. Nor is the free reign of the market liable to erode the recognized factors that impede adequate allocation of resources to efficiency and conservation, such as the limited likelihood of those who privately pay for and install measures accruing all the subsequent benefits. As McGowan (1987) argues:

> Privatisation of the electricity supply industry . . . can provide incentives for energy efficiency. But if the opportunities for improving efficiency are to be fully taken up, it will require a more robust role for government than is currently exercised.

In short, UK expenditure was already deficient in this area and, in embarking upon privatization, this tends to place upon the state an even greater need to make up for the failures of private provision. Yet, the same market philosophy that makes privatization imperative also leads to a withdrawal of state expenditure in support of the private sector. For no sooner had the bill for privatizing electricity been published than the Government had announced a cut in the budget of the Energy Efficiency Office from £24.5 million to £15 million. As the Energy Select Committee (1981/82: 16) observed:

> The Steering Committee is well aware that it is not the Government's policy to provide fixed or other incentives for energy saving. It believes, however, that there is a case for reviewing this policy.

As privatization drew nearer, the NEDC (1986: 3) similarly had limited faith in the efficacy of the private sector:

> The overall targets being set by the Government for energy savings are unlikely to be achieved as the result of current market forces.

This merely emphasizes the continuing relevance of the classic criticism of governmental neglect of conservation in contrast to its fanaticism for nuclear power (Energy Select Committee 1980/81: para 32):

> We were dismayed to find that, seven years after the major oil price increases, the Department of Energy has no clear idea of whether investing around £1,300 million in a single nuclear plant (or a smaller but still important amount in a fossil fuel station) is as cost effective as spending a similar sum to promote energy conservation.

Similar considerations apply to the development of Combined Heat and Power. This is a system of generating power that converts less fuel into electricity but is more thermally efficient in that otherwise wasted heat in flue gases and cooling towers is used to provide space heating and hot water. As an energy system, it requires investment in and co-ordination of electricity supply with surrounding buildings that are to use the non-electrical forms of heating. The UK is the most laggardly European country in the use of CHP. Yet, as the Combined Heat and Power Association pointed out in its Memorandum to the Energy Select Committee (1987/88a, Mem. 63: 203), 'the White Paper makes no clear commitment to energy efficiency and fails even to make a passing reference to Combined Heat and Power.'[17]

These observations concerning the neglect of conservation, energy efficiency, and alternative methods of generating electricity, of which CHP is just one amongst other methods, point to an entrenched institutional system of supply which will be reinforced by privatization. This is confirmed by Thomas' (1988) analysis of the opportunities offered by the refurbishment of existing plant rather than the building of new. He sees this as increasingly economic because of four factors; slow-down in technical progress in power plant performance, reduced and uncertain demand prospects, an end to the fall in the real cost of producing electricity, and the problems on large-scale construction sites.

In principle, of course, refurbishment would be undertaken whenever economic to do so, especially in the context of a privatized industry. But it is far from clear that the two incumbent power station companies will have the regulatory incentive to refurbish, and they will presumably have the power to prevent new entrants from employing their old stations. More generally, the prospects of any competition at all from new entrants is extremely remote. As Power in Europe (1988) puts it:

> the Bill does very little to explain the fundamentals of the privatized system, in its daily operations, or in key matters such as the pricing formula. These will appear through the contracts and licences, which should surface in the course of 1989 . . . (the pricing of) the industry has to be arranged on the Big Fix principle; otherwise it is unsellable.

The 'Big Fix' principle is not, however, confined to the issue of the future relations between the two companies. The Government announced along with its privatization plans a two-stage increase in electricity prices by 15 per cent in order to fund a future investment programme of £40 billion. Those committed to *laissez faire* should have preferred reliance upon the private capital market to generate such funds. Further, the Government is taking upon itself the role of predicting and making funds available for future investment through the price increases. In short, the price increases are much more transparently a device for raising the profitability of the industry in the future in order to guarantees the success of the privatization sale in the present.

More generally, Hope (1987) has argued that an estimated central value of the CEGB's generating capacity in terms of present value is approximately £18.5 billion. But this is extremely sensitive (by as much as £5 billion) to three factors over which the Government will retain continuing influence – the rate of retirement of old nuclear power stations, the cost of coal, and the price to be charged for electricity. Clearly, policy over these is already being fixed in favour of profitability of electricity supply.

Perhaps the most important Big Fix will be the one of the past that became known as the 'Buggins' principle whereby the builders of power stations took turns in receiving orders from the British government – Buggin's turn next.[18] Power station construction has long been recognized to have been unsatisfactory in the UK with substantial cost over-runs and delays in completion. This has received little more than token attention in the debate over privatization. It is simply presumed that privatization will lead to a significant improvement through the tougher contracting of generators motivated by profit and, otherwise, through the increased international competition that is made available by not relying exclusively on domestic producers, especially with the coming of the Single European Market of 1992.[19]

Such an approach is simply to gloss over the problems involved. They include the cartelization of the industry at a global as well as at a domestic level against which neither the Single Market nor tougher contracting will be effective in and of themselves. There is also the nature of the British construction industry to consider (see Ball 1988). It has a fragmented structure, a casualized workforce, and a dependence upon a multi-layered system of subcontracting. This has much to do with poor industrial relations and management and with the variable capacity and willingness to complete or delay work as demand on individual subcontracted companies is highly volatile. Nor has the case for improved power station supply through private contracting been considered, let alone established, by scrutiny of comparative performance in large-scale for the private as opposed to the public sector.

This is not to say that the power station supply industry is unchanging. Currently, it is going through a dramatic transformation, with mergers, acquisitions, and joint ventures proceeding at a pace that has been described as being as rapid in the past two years as in the previous fifty. Now six or seven giant companies are set to dominate the world scene, one of them being around GEC which has come to terms with Alsthom of France, General Electric and Combustion Engineering of the United States. Two points need to be made about this. First, such developments are independent of the process of privatization, as they are in part inspired by continuing economies of scale in power station construction, as documented by Cecchini (1988). Second, smaller and privatized generating companies are liable to have a harder time benefiting from the resulting economies of scale and combatting cartelization than a single state-owned utility. Developing countries, for example, are known to pay over the odds for electrical equipment to the internationally cartelized suppliers.

There is also the prospect of such power station producers diversifying into generation and providing themselves with guaranteed contracts. The Government has recognized this but only through the nominal gesture of preventing any single shareholder from owning more than 15 per cent of any distribution company in the first five years. For the five years after that, this restriction will fall with the consent of 75 per cent of shareholders, and all restrictions will then be removed. Equally diversification could be downstream into fuel supply, so that a single company could supply coal, build power stations, and generate electricity. The scope for monopoly abuse is wide as are the avenues for creating it. It is worth recalling that Rockefeller's monopoly of the nineteenth century American oil arose out of the gaining of control of the distribution system.

The restructuring of power station supply is part and parcel of a much wider reorganization amongst those companies that are engaged in the electrical and electronic sectors. At least four separate major industries are involved – defence, electrical consumer durables, telecommunications, as well as power engineering. It follows that industrial policy for any one of these sectors may be made on the basis of merger activity around any of the others. This process is at its sharpest when one conglomerate acquires another and sheds or strips the productive assets for which it has no use and/or excess capacity. It stretches the economic imagination to believe that this is the most advantageous way in which to make industrial policy. Yet, in the case of the merger battle between GEC and Plessey, the central issue is around telecommunications, as far as these two companies are concerned – although government policy may be made out of concern for sovereignty in defence provision – but it is all four sectors mentioned above that will inadvertently feel the effects.

This all has its counterpart in the arena of R&D, previously raised in the last section in the context of nuclear power. As Holmes *et al.* (1987) observe, electricity supply is the largest single UK industry and employer and at the heart of the development of computing, telecommunications, and new technology generally. Whatever its own contribution to R&D directly, it has a profound effect upon it through its purchases alone.[21] Ince (1988) notes that the privatization White Papers make no reference to electricity as an industrial customer. Yet it is a major customer of manufacturing – from nuclear power stations to hand tools – as well as the western world's largest purchaser of coal, the showcase for UK power plant exports, and the proxy customer for UK mining equipment. For these, there are economic multipliers as well as R&D spin-offs. But the standard criticism, for example, of UK R&D policy in the area of information technology is that there is a lack of an overall coherent and co-ordinated strategy.[21] This is only going to be consolidated by privatization.

In short, this section has revealed a number of defiencies with the privatization programme for electricity. These are related as much to continuities as to breaks with policies of the past. They do concern issues of static efficiency but equally ones of dynamic change, as in industrial restructuring and R&D. The withdrawal of state intervention in the form of public ownership is only going to consolidate existing weaknesses in the British economy, whether it be in providing the agency for energy efficiency and conservation measures or in co-ordinating industrial development across sectors. These points have already been demonstrated in the previous section in the context of nuclear power. The same will now be seen to apply to the coal industry.

Coal and 'uneconomics'

During the course of the miners' strike of 1984/5, the central issue appeared to concern the closure of so-called 'uneconomic' pits. In these terms, British Coal and the Government were to be seen as representing the hard face of economic realities as against the miners whose case for sympathy rested on the saving of communities and jobs. The very framing of the issue in these terms leaves it open to contain an implicit answer in favour of closure. Uneconomic, after all, is uneconomic, Cumberbatch *et al*. (1986: 89), quoting Mrs. Thatcher:

> There are an awful lot of uneconomic pits and you don't need to argue about the definition. They are heavily loss making pits . . . . You don't need to argue about them . . . . You have to go through a procedure with the NUM and they have to be shut down.

In fact, the procedure with the NUM is supposed to reassess a closure decision and, in principle, allow the pit to remain open. Leaving this aside and given the centrality of such 'uneconomics', it is peculiar that very little discussion and even less quantification of its dimensions emerged during the dispute. It might be expected that the presumed cost of maintaining uneconomic pits open would be carefully calculated so that a choice could be made whether jobs and a community or two or three were worth saving even at the expense of being uneconomic. To a large extent, such a choice was already rejected by virtue of the Government's monetarism, for which the free market is idealized as the economic arbiter of commercial viability (even if the

Government has been a far from neutral and inactive bystander).

What stood out during the strike was the relative absence of the discussion of these questions (see Saville 1986). To some extent this is explained by the focus on the conduct of the strike – whether it be the initial issue of a national ballot or the continuing problem of picket line violence – rather than its immediate substantive causes.[22] But two further factors are significant which point to the reluctance of both the Government and British Coal to engage in debate over economics. Polemically, Scargill's and hence the NUM's position was always presented in the ludicrous form of no pit closure short of complete exhaustion, whereas the miners offered the 1974 Plan for Coal as a basis for negotiations (under which seventy pits or so had already been closed in the decade before the strike). Analytically, criticisms of the uneconomics of pit closures had been made. They received some limited publicity through the media and the NUM but remained unanswered. In short, whilst the uneconomic was the ideological terrain on which British Coal and the Government chose to engage the miners, it was an analytical terrain that was left obscured.

This is not surprising in view of the powerful arguments against pit closures. For the economy as a whole, the costs of a closure programme can be measured in terms of lost employment and output. It is relatively simple to see that a miner in work producing coal is liable to be more economic than one on the dole. But the matter is more complicated than this because a closure programme affects different economic agents unequally and in different ways. An unemployed miner receives redundancy payments and the dole but loses wages. The Government has to make the payments to the unemployed miner and loses income tax and national insurance payments. Other workers supplying goods to British Coal will lose their jobs, etc.

Using calculations such as these, Glyn (1984) found that over a seven-year period the government would save about £7,000 in subsidy per laid-off worker per annum but, for a 57-year-old worker taking redundancy, would lose almost twice as much – in losses which would not figure in the British Coal accounts since they were borne by government. More generally, taking the 12 per cent of highest cost pits, closure would lead to the loss of 40,000 miners directly and 35,000 further jobs from other industries supplying British Coal or depending on expenditure of wage revenue. Whilst £275 million would be saved in subsidy to British Coal, £475 million worth of coal would not be produced and £480 million would be lost in tax revenue and unemployment benefit.

The exact figures in these calculations are not so important as the fact that the calculations can be made and should be by any policy maker implementing a programme of closure. But these calculations depend on two important and closely connected assumptions; that the coal that would be produced in the absence of closures could be usefully employed and that the miners made redundant could not find alternative employment (thereby releasing the need to make dole payments). These assumptions are inconsistent for an increased use of coal requires fuller employment, either in industry where it is used or to provide the incomes of those buying the coal (although income redistribution, e.g. free coal to pensioners, would also provide an attractive 'market' for some coal without higher employment other than in the coal industry). So, if there were a market for the coal, this must mean fuller employment, this would mean alternative jobs for miners and no need to pay their dole, thereby reducing the cost of closures. At one stage, Glyn justifies his calculation of the cost of closures on the need to pay the dole by referring to 'the persistence of mass unemployment' whilst, at a later stage, he argues that 'the reason that coal sales are so low is that production and incomes are so depressed.' But, if this were otherwise, there would, of course, be greater job opportunities for displaced miners and some pits could become uneconomic because allowance for dole payments would be unjustified.

This inconsistency in Glyn's approach is not serious. It merely points to the major organizing principle of his argument – that it is better to have full(er) employment at higher levels of output for which more and more (coal) production becomes economic. On this basis, he is able to show that each and every colliery in the UK is economic in the sense that there are net social costs from closing even the least profitable pit – Celynen South with a net loss of £59.5 per tonne turns over a net social gain of £1.9 million from being kept open. Adding up over all of the then 164 pits for which the figures were available leaves a net social gain of £2,783.0 million rather than the loss of £181.8 million representing British Coal's deficit.

Glyn is able to make these calculations for the individual worker, for the individual pit, and for the industry as a whole because the principle being applied is not specific to the coal industry (see O'Donnell 1985). It is rather a general case for employment, made with application to the coal industry. It could equally apply to the steel or car industry and not necessarily to a nationalized industry.[24] Closures of private firms or their plants would warrant subsidy to losses, whether or not accepting justifiable arguments for an equity stake in return for such support.

This does not invalidate Glyn's position but merely points to its strengths and its weaknesses. Whilst an important intervention on behalf of the miners during the course of the strike, it is better seen as part of a strategy for higher levels of employment in which each sector of the economy is expanded. Otherwise, Glyn's economic case for the coal industry would be reduced to a special pleading for coal, as was done for wages in the early 1970s, for the case would apply equally to every other sector, plant, or unemployed worker, although the exact figures would be different in each case (and subsidy would be most effective in job creation in the most labour-intensive industries, taking account of direct and indirect effects).

A specific case for coal can be made in terms of its cheaper thermal cost. It provides more energy per unit cost than any other conventional fuel. This is dramatically illustrated by the costs of the strike itself which witnessed substantial substitution of oil for coal in the generation of electricity. The extra cost of generating electricity during the course of the strike was of the order of £1.5 billion – compared to total normal oil and coal cost of £4.3 billion (GLC 1984a). This reflects the need to switch from larger more efficient coal-fired to smaller oil-fired stations, as well as the previously mentioned cheaper thermal cost of coal – gas is 18 per cent more expensive, oil 97 per cent, and electricity 163 per cent (Labour Research 1984).

Nor is it legitimate to accept the case for coal but seek to rely on imports rather than domestic production. The spot market for coal does have a low price due to the excessively depressed markets for energy. It is generally recognized that all but possibly South African and some Australian producers are unable even to cover working expenses let alone fixed capital costs.[25] So long-run prices are liable to rise and, even so, it is erroneous to base economic calculations on a presumed international price of coal, a market on which only a small, if growing proportion of coal is bought and sold (less than 7 per cent of world consumption in 1985). As Labour Research (1984) commented, 'Using world coal prices as a criterion is like fixing bus fares on the basis of international air fares.'

This raises the question of what is the appropriate price to use in evaluating the economics of the coal industry. It is important to recognize that the market price, even if it were appropriate in principle, is not available in practice. With as much as 70 per cent of coal produced in the UK being consumed by the CEGB, the price set simply reflects a transfer of income from one nationalized industry to another. As Kerevan and Saville (1985a: 9) propose:

> a change in the price structure of coal which will allow the NCB to cover its genuine operating costs and meet its capital requirements . . . altering the transfer pricing arrangement between the two nationalised energy groups . . . to divert the excessive profits of electricity derived from coal back into maintaining coal capacity.[26]

Kerevan and Saville also argue for rationalization of the debt structure of British Coal. This raises the more general question of the extent to which its accounts reflect its economic performance and the extent to which the accounts are cooked to instigate a strategy for closures or an individual pit closure. Glyn points to the fact that losses ascribed to British Coal have not always been incurred by them and, in any case, do not enter into future costs of production and so would not be saved by a closure programme. This includes compensation for surface damage (resulting from past mining whether by British Coal or not), pension payments for past employees (rather than contributions for those currently in employment), and social costs for redundancies made in the past. All of these would have to be made irrespective of British Coal's continuing production programme. They amounted to an annual burden of almost £700 million. To consider these as costs of the industry is like charging private employers for the dole and pensions of their previous employees, in addition to their normal national insurance contributions.[27]

In addition, there is the payment of £467 million in interest of which £400 million is a straight transfer to the Government as the source of the loans. These payments are not costs of production but represent an accounting device that would only have a rationale for decision making if the rate of interest reflected the cost of capital used that could otherwise be placed in an alternative sector. Such an ideal capital market is far removed from the realities of a highly variable real interest rate and a low level of capacity utilization. In any case, Glyn suggests that British Coal pays interest at twice the rate of other nationalized industries and half as much again as the dividends, etc. of private companies. It must also be reckoned that the need to borrow depends on the past history of internal funds generated which have themselves been limited by pricing and other policy towards the industry.

These factors are brought forward to answer the charge that the operations of British Coal as a whole are uneconomic because of its deficit and the need for a government subsidy of £1.3 billion in 1983/4. The conclusion reached by Glyn by reconstructing the accounts is the opposite:

> Sales of coal, even in the present depressed conditions, are practically enough to pay the miners and for other inputs actually used to produce the coal including a proper level of depreciation. The general state of the industry's finances cannot be used to justify in any way the NCB's pit closure programme.

But the more general point is that the accounts, whether reformed or not, remain a poor basis for rational decision making or, to be more conspiracy-minded, are more a device for the justification of decisions already made. A similar point arises for the pricing of coal in the relations between British Coal and the CEGB. Here, there is a case for organizing the two as a single energy utility for which a co-ordinated plan of coal use and power plant construction can be implemented.

Privatization and property rights

This discussion suggests that those pits labelled as uneconomic may well be worth keeping open. But this leaves unexplained why some pits rather than others become designated as uneconomic in whatever sense. Here, two factors are involved. The first is the explanation of the varying performance of individual pits. Why are some more productive than others? Or, more exactly, why do some appear more productive than others in British Coal's figures? For, the second factor is the way in which varying performances are represented. To what extent do the profit and loss figures of individual pits reflect their relative performance?

In the first instance, the traditional wisdom is that geological conditions are the main determinant of pit productivity. Collieries can be listed in a descending order of fertility and, as you move down the league table to satisfy demand, you cut off the pits in the 'relegation zone' once demand has been satisfied. It has already been suggested that demand is highly dependent upon the overall level of economic activity and the balance of fuel usage, in terms of the sorts of arguments put forward by Glyn. Here, there is the irony of self-fulfilling prophecy in the person of Ian MacGregor who appears to have moved around the input–output table of the British economy, a kamikaze personification of the Keynesian multiplier in downward spiral. As a member of the Board of British Leyland, he saw its production drastically reduced. Car manufacturing is a major steel user. As Chairman of British Steel, MacGregor also witnessed a dramatic decline. Whilst, in the 1970s, it was planned that steel output should rise by 50 per cent, it has fallen by that amount. Associated with this is a fall in the demand for coking coal, equivalent to the loss of ten million tonnes' output to British Coal – MacGregor's final port of call in his management trip around the nationalized industries (see Beynon *et al.* 1986).

But demand aside, can the league table of collieries, on which it supposedly exerts its cutting edge, be legitimately identified? It is crucial to recognize that practically no public evidence has been brought forward to support the view that the problem of high-cost pits is one of inevitable and unavoidable geological conditions. It is simply inferred to be so and this is very convenient for a closure programme. For opposition can then be seen as a futile struggle against nature itself. In the Monopolies and Mergers Report of 1983, it was reported that productivity changes between 1950 and 1975 were explained by the following factors: 30 per cent by face mechanization, 28 per cent by powered supports, 13 per cent by length of face, 19 per cent by seam thickness, and a mere 10 per cent by regional distribution of production. Even if these factors are not independent of each other (extra seam thickness may be regionally determined, for example), nonetheless, the most important determinants of productivity appear to come from improving technology.

This is confirmed by the Monopolies and Mergers Commission (1989) for the five-year period straddling the strike during which output per manshift increased from 2.44 to 3.62 tonnes. Whilst total output declined by 20 per cent, manpower was almost halved to just over 100,000. Production was concentrated on 246 faces (reduced from 574) and in 94 collieries (down from 191). However, the report argues that three-quarters of the productivity increase came from the use of heavy duty supports and other investment and modified working practices. Consequently, concentration of production on fewer faces in even fewer pits does not appear to have been the major source of productivity increase (although the various factors listed above are not independent of each other). In other words, the closure of uneconomic pits was mainly concerned with reducing output, and these pits may well have achieved comparable results to those retained if they had received the same level of heavy duty investment.

This suggests that it is not the natural inferiority of the so-called peripheral coalfields that makes them high cost relative to the central core but the unfavourable differential allocation of investment funds which has deprived the periphery of the most advanced production methods. This is borne out by Figure 9.1 which illustrates the close connection between low investments and high costs across the different mining areas.

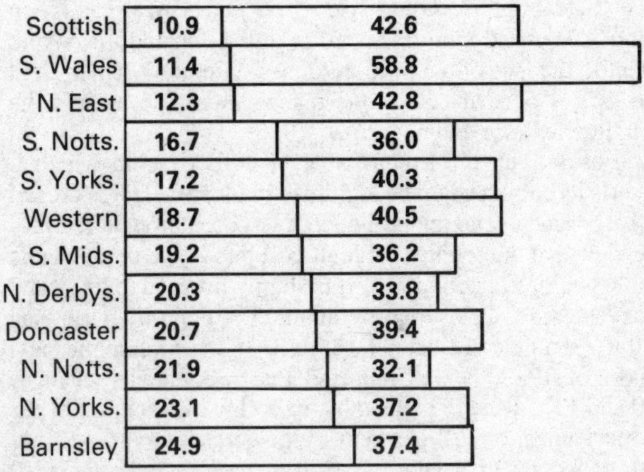

Area	Investment	Operating cost
Scottish	10.9	42.6
S. Wales	11.4	58.8
N. East	12.3	42.8
S. Notts.	16.7	36.0
S. Yorks.	17.2	40.3
Western	18.7	40.5
S. Mids.	19.2	36.2
N. Derbys.	20.3	33.8
Doncaster	20.7	39.4
N. Notts.	21.9	32.1
N. Yorks.	23.1	37.2
Barnsley	24.9	37.4

Figure 9.1 Investment per head (£000s) and operating costs £ per tonne by NCB area
Source: *Labour Research*, September 1983

In addition, there is a high variability of pit performance, up and down, year in and year out, for high- and low-cost pits. Productivity has moved annually by as much as 16 per cent on average. This makes the identification of the uneconomic pits highly problematical – when at least 50 per cent of the time, the productivity difference between any two pits will change by a third in one year. Unsurprisingly, pit performance can be turned around by a major investment which lifts it out of the closure zone (see O'Donnell 1988).[28]

Whatever the productivity performance of the different pits, their results still have to be translated into commercial terms in an accounting sense, and this too has been the subject of dispute.[29] For cost, there is, for example, the problem of assigning fixed overheads for the industry as a whole to individual areas and collieries. To a large extent, these costs, which constitute as much as a quarter of operating costs, are not reduced by individual pit closures and so should not figure in the decision on the economics of an individual pit. There are also considerable doubts, in any case, about the procedures for dividing such costs between areas and pits.

In terms of revenue, this depends heavily on both the level of output for a pit and the price assigned for that output. In conditions of excess supply, British Coal assigned a much lower price for coal to pits that exceeded their quota. This created a disincentive to produce beyond full capacity. On the other hand, a higher level of output quota means that overheads, justified or not, can be spread, thereby reducing average costs. Clearly, individual pit performance will depend crucially on the assignation of quotas which are not made in any formalized fashion. As a result, however, the relative performance of pits was heavily influenced by quota allocation and could be the mechanism, conscious or otherwise, for moving one pit rather than another into the relegation zone. As Berry *et al.* (1985a) conclude:

> Assessments of colliery performance thus run the considerable risk of becoming self-fulfilling prophesies; alleged high performance yielding high investment and output allocations, both of which contribute to low unit costs. The opposite consequences occur for so-called low performing collieries.

In short, whether in the allocation of investment and output or in the construction of accounts, British Coal tends to create uneconomic pits for closure rather than unwillingly responding to their inevitable emergence. What are already self-fulfiling practices are further consolidated by the short-term operation of those pits on the closure list. With lack of development work, average fixed costs escalate with lower output, workforce morale slumps with knowledge of this and redundancy, transfer, or absenteeism become preferred options. A lower pace of extraction from the continuing faces can even prolong the life of the pit and the jobs associated with it. Kerevan and Saville (1985a: 46) describe the situation:

> At Cardowan, as in a number of other pits, a long-term decision by Scottish Area NCB to close the pit resulted in a slow reduction in capital investment in drivages. It was clear by the end of financial year 1982/3 that no further capital expenditure on drivages or anything else was earmarked for Cardowan. Such uncertainty, coupled with other local industrial relations problems, produced a high rate of absenteeism. This obviously affected short-term OMS. A management decision to run-down the pit produced the low OMS which justified the action.

Here, there is a parallel with the role of accounts in the commitment to nuclear power, as discussed in the first section. Initially, the policy is already determined so the accounts are treated casually as long as they are supportive of what has been decided. Once, subject to criticism, they are reconstructed with greater care but with the result unchanged. This is not to argue that accounts have played no role, nor that pit have been selected for closure randomly. Rather, the whole issue of 'uneconomics' has a fantastic element to it, part horror and part imagination, like the 'undead' of the movie genre. Once the misconceptions in which it is shrouded are cast aside, then the issue of colliery closures can be more adequately considered in a wider and more realistic economic and social framework.

The strike of 1984/5

The miners' strike of 1984/5 has been subject to a number of different analyses and characterizations, as in Adeney and Lloyd (1986), Callinicos and Simons (1985), Crick (1985), Goodman (1985), MacGregor (1986), Samuel *et al.* (1986), and Sunday Times Insight (1985), each reflecting different approaches as well as different sympathies, although Winterton and Winterton (1989) is the most informative account – see also Brown and Selzer (1985 and 1986) for press coverage and Green (1985) for a bibliography. Accordingly, Gibbon (1988) has critically surveyed the literature dealing with the strike by pointing to a number of themes that have been taken as central – the break with consensus corporatism, an energy establishment and/or governmental conspiracy to break the miners and/or labour movement, a means of forcing through new technology, the vanguard role played by the NUM, or Scargill personally, in labour movement politics, the defence of mining communities' way of life, the bureaucratic union leaders holding back rank and file militants or radical leaders dragging unwilling members into conflict, and the conflict between national and local union organization, the significance of women's involvement, and so on. Gibbon's own preferred approach is to emphasize the conflict over the right to manage, although his contribution neglects the economic issues discussed in the previous section.

The complexity and range of issues involved points to the significance of the strike. Closures had occurred before without conflict; closures and closure programmes had equally been withdrawn. This suggests that the strike had much more at stake even if this was ratcheted upwards by the severity and length of the conflict itself. One way in which this can be discussed is in terms of the exercising of property rights. Much was made by the Government of the right of miners not on strike to work. Ian MacGregor (1986) discusses the strike as a strategy to re-assert the right to manage. In this way, the dispute assumes a struggle between those who seek to preserve the rights of property, either of labour or of capital, even at the expense of the conflict involved. A more sophisticated approach, however, recognizes that property rights are not so simply defined, and that the dispute revealed the extent to which they are intimately bound to wider economic and other social relations, as will emerge during the course of this section. In retrospect, it can also be argued that the different conflicts raised by the strike not only exposed the many dimensions of property rights but also acted to transform their content. For, at the end of the day, the miners returned to work facing, first of all, a renewed closure programme and, secondly, the eventual promise of privatization.

Within Marxist theory, capitalism is specified as a mode of production in which the wage labourer is free to sell labour power, not bound to any single employer and equally able to withdraw labour, even if at the expense of income loss. But it must be recognized that this is an abstraction and that the nature and extent of these freedoms, like others in the political domain, have to be won and defended. After all, miners in the Scottish coalfields were annually bound to their employers as late as the beginning of the nineteenth century. Nor is this simply a matter of improving the rights and quality of life of the working class. It has a profound effect on the nature of capitalist development, Marx arguing in Volume I of *Capital*, for instance, that limits to the length of the working day act as a great stimulus to the introduction of machinery. But also in the context of the length of the working day, Marx (1965: 235) recognized that, in principle, the capitalist has the right to command labour through ownership of the means of production just as the labourer has the right to withdraw work as the formally independent owner of the commodity labour power.

The coal question

'There is here, therefore, an antinomy, right against right, both equally bearing the seal of the law of exchanges. Between equal rights force decides.' It follows that force is an economic factor, emerging explicitly when mutual agreement concerning terms of exchange over property rights breaks down.

This is the context within which the course of the miners' strike can be viewed. It was on the part of the employers, an alliance between British Coal and the Government, a concerted and ultimately successful attempt to prevent the miners from exerting their collective right to withdraw their labour as a means to defend their jobs in the longer term. The preparations of the Government were both well-established and strengthened by the course of events over the previous decade. At the end of 1977, productivity deals were introduced in the industry and these had the effect of weakening unity across the coalfields. This was reinforced by developments under the Plan for Coal, agreed by the NUM, British Coal, and Labour Government in 1974 and including an output target of at least 135 million tonnes by 1985. The Plan was endorsed with some amendment as late as 1981 by the Thatcher Government. But it was against a background of declining production so that by the time of the dispute, deep-mined output had fallen below a hundred million tonnes per annum. The Plan led to the allocation of heavy investment to the central coalfields, guaranteeing greater security of employment and higher earnings there whilst threatening the peripheral coalfields when energy and coal demand fell far short of expectations.

In 1978, whilst still in opposition, a plan to challenge the power of the unions, learning lessons from the defeats of 1972 and 1974, was drawn up for the Tories by Nicholas Ridley and leaked in the *Economist* (27.5.78). It planned to isolate and defeat a public sector union through weakening the rights and ability to picket and by denying welfare benefits to strikers. The Thatcher Government of 1979 enacted the legislation to back up the implementation of the Ridley Plan both in trade union legislation and in welfare rights of those on strike. The Employment Act of 1980 was aimed at preventing secondary picketing, whilst the Social Security Acts denied benefit to unmarried strikers and presumed strike pay of £16 per week for those with dependents, and this was deducted from benefit payable.[30] For the Tory Ministers, 'There is the choice not to strike, to go back to work and to earn the living that is available', whilst the denial or reduction of benefit to strikers 'will save public money to a modest extent, but that is not its main concern. The government was elected . . . to restore a fairer bargaining balance between employers and trade unions' (Jones and Novak 1985: 92).

A tentative move towards confrontation was made in February 1981

Privatization and property rights

with the announcement of a twenty-three pit closure programme. This was soon withdrawn against the threat of national industrial action. No doubt the Government felt that they could not be sure to win a strike at this time. But within three years preparations had been consolidated. Coal stocks had been increased from 37 to 57 million tonnes and the climate of industrial relations had been tested at disputes around the NGA (including police violence against pickets) and GCHQ. Moreover, the Tories had been re-elected with a massive majority in the wake of the popularity surrounding the Falklands venture; a reputation for a firm hand in winning the last election coincided with a long interval until the next.

The greatest encouragement, however, to the Government came from other developments within the industry and the union. Between March 1981 and March 1984, 41,000 jobs were lost. By as early as July 1982, 15 of the 23 pits listed in the officially denied 1981 'hit-list' had been closed or merged.[31] More significantly, there had been votes against strike action in November 1982 and March 1983. The particularly severe incidence of the closure programme on Scotland and South Wales had failed to prompt adequate solidarity from the other coalfields. Against this background of weakness, an overtime ban was introduced in October 1983 to reduce coal stocks and to highlight the low level of basic pay. But when British Coal provocatively announced the closure of Cortonwood in breach of agreed procedure, and in the area with the most resolute commitment to defend jobs, the NUM was ill-prepared for the confrontation ahead.

This is reflected in successive extracts from the *Financial Times* for 1984:

9 March	Odds Stacked Against the Miners
10 March	Miners Strike Action Will Be Patchy
12 March	S. Wales Miners Reject Strike
13 March	Flying Pickets Go Into Action As Strike Halts 50% Of Pits
14 March	More Pits Close As Pickets Step Up Pressure

By 15 March, it was reported that 132 pits were idle out of a total of 174. What this confirms, apart from the lack of worth of yesterday's news, is the extent to which the NUM was ill-prepared for the strike ahead. Individual areas were in favour of strike action when they felt the threat of closures, but this had operated to create cumulative disunity in the wake of insufficient support for national action.[32]

Whatever the unified commitment to strike action in advance, the traditional method for achieving it subsequently by picketing had been anticipated by the police. Measures were put in hand in 1972 after the coal strike then had revealed the effectiveness of flying pickets. The National Reporting Centre (NRC) was set up under the management of the Association of Chief Police Officers (ACPO) (see Kettle 1985a). Its function is to monitor the availability of, and need for, police reinforcements across the nominally independent local police forces. Although the NRC, in principle, is not supposed to be an operational centre in the sense of imposing orders on local police forces, in practice, it is inevitable that such monitoring acts heavily to create and co-ordinate a national policing strategy when the NRC is in full operation.[33] So, from 1972 on, new forms of policing were available in the event of widespread attempts at picketing.

The police strategy first involved the prevention of pickets from moving by turning them around at road-blocks. The legality of such policing was far from clear and, on the ground, was certainly unknown to the police officers themselves. In the first six months of the strike, it is estimated that 290,000 pickets were impeded by the police on the grounds of being liable to cause a breach of the peace hundred of miles away from their ultimate destinations (Blake 1985). Those that were able to picket or intending to do so were subject to arrest and harassment whether on the picket lines or in their communities, not under any new trade union law, but under long-standing criminal law enabling the police to arrest at will. Over the whole dispute there were in all just under 10,000 criminal charges brought during the strike, of which over 40 per cent comprised 'conduct conducive to a breach of the peace (Section 5 of the Public Order Act 1936)' and only a minority involved serious violence (about three or four per day on average).[34] As McIlroy (1985: 81) concluded:[35]

> almost any effective picketing involves the commission of one or more criminal offences . . . its conduct almost entirely in the discretion of the police.

Similar discretion was applied in the courts, again even to the extent of traversing the boundaries of legality (see Christian 1985). Those charged were released on conditional bail, uniformly applied and subverting the principle of justice for the individual, with the intention of preventing renewal of picketing. Those for whom charges were unlikely to stick were bound over to keep the peace, that is not to picket.

It has already been mentioned how welfare legislation limited the benefits available to striking miners. Sources of support were also restricted by the policing of street collections through harassment and arrest for obstruction and under the 1824 Vagrancy Act.[36]

Such policing was completely independent of the trade union legislation newly introduced by the Tories which played a lesser but, nonetheless, significant role. Within a few days of the strike beginning, British Coal took out an injunction against the NUM to prevent secondary picketing but it was never served. Rather other private companies and working miners took civil action against the NUM leading to sequestration of assets and, otherwise, sterilization of union funds that had been moved abroad.

A major influence of the trade union legislation was, however, ideological for policing and the criminal law have usually proved more effective in practice in dealing with industrial conflict.[37] But changing law signals to the public as well as to magistrates that enough is enough and that unions and union activity are to be seen as getting out of hand. This, in turn, supports what has been a standard practice of policing – to criminalize a section of the population by identifying it as collectively carrying the characteristics of guilt. The most recent example is the criminalization of the black population (whether as muggers or as illegal immigrants),[38] but the same approach applied to striking miners. They were liable to picket, to break the law, and consequently belonged to the criminal classes and were subject to the harshest conditions that could be applied by the police and the courts.

Support for such an ideology could not rest on policing alone but dovetailed with the Government's propaganda, ably supported by the media. The issue of a national ballot formed a major starting point but it gave way to a complex and inconsistent configuration of arguments. Whilst the Government demanded the right of management to manage, it also demanded the right of workers to work – with hundreds of police at times symbolically enforcing that right for a handful of scabs. But the right of management to manage contradicted the right of workers to work, those threatened by the closure programme. The Government also demanded the rule of law and order.[39] Yet, at the same time, it oversaw a massive police operation that witnessed police violence, provocation, and other violations of the law.[40] Finally, the Government claimed that the strike threatened parliamentary democracy. Yet, it also wished, at least in principle, to argue that the resolution of the dispute, if not the terms on which it was conducted, did not concern the Government at all but was an issue between employer and employee.

Of course, the principles and practices of the Government bore very little relation to each other, nor to consistency and reason, except in their mutual objective of defeating the strike. A measure of their failure at the ideological level was the attempt to criminalize the NUM as 'the enemy within', in Thatcher's phrase, bringing the Falklands war to mainland Britain from the Malvinas:

> At one end of the spectrum are the terrorist gangs within our borders and the terrorist states which finance and arm them. At the other

are the hard left operating inside our system, conspiring to use union power and the apparatus of local government to break, defy and subvert laws.

(Mrs Thatcher, quoted in Wade 1985: 281)

The ideology of the 'enemy within' was soon dropped, as it proved unpopular so substantial was support for the miners,[41] but the police and courts continued to enforce their designated interpretation of property rights regardless, with the full support of the media.[42]

Concluding remarks

In its early months, it became commonplace to compare the strike with those of the 1970s and with the General Strike of 1926, although the latter became increasingly popular as defeat approached.[43] In the latter case, the points of similarity are certainly attractive and they provide a cutting edge with which to examine the situation in the 1980s. Scargill stands comparison with A. J. Cook. Divisions within the union were reproduced with the miners continuing to work in the central coalfields in the 1980s, whilst the 1920s witnessed the rise of Spencerism in Nottinghamshire, but also in South Wales. Both strikes suffered from the weakness of support from the trade union movement and the leadership of the Labour Party. Both were the culmination of a series of conflicts in some of which the Government had been defeated but through which it became increasingly determined and prepared to inflict a defeat.

This by no means exhausts the parallels but it is as well to be aware of the dangers of the exercise. The most serious is in reading 'what happens next' by reference to what happened before. The point is that the future has yet to be made and references to the General Strike and its aftermath as a precedent are liable to be defeatist as far as the miners and the Labour movement are concerned. A notable exception to such defeatism is provided by Kettle (1985b) but only because, to exaggerate, he transforms the General Strike into something of a victory for the working class. He does so by reference to the election of a Labour Government within three years and the supposed creation of trade union influence over economic policy through the Mond–Turner talks.

The re-election of Mrs Thatcher in 1987 has already rendered this view obsolete. But it also takes no account of the failures of the Ramsay MacDonald Government and its ignominious collapse into the National Government of 1931. Nor does it properly assess the status of Mondism. As has been amply demonstrated, the capitalists in these negotiations predominantly represented the interest of large-scale capital seeking an alliance with labour for schemes for the rationalization of industry. (See McDonald and Gospel 1973 and Dintenfass 1984.) These, certainly through the talks, led to nothing. For whatever Mond and Turner could agree met with resistance from smaller-scale capitalists and prevented Government intervention, however much sympathetic. As seen in previous chapters, for the coal industry itself, rationalizing schemes through amalgamations were obstructed in the 1930s through the opposition of entrepreneurs.

This is a signal of major differences between the 1920s and the 1980s, most significantly within the industry itself. In 1926, the strike directly concerned wages and hours, although it also reflected long-standing grievances over safety and welfare and other such considerations. An increasingly important solution to the industry's problem was seen by the miners in terms of nationalization. By contrast, the dispute of the 1980s revolved around closures. With the industry already in public ownership, the miners' case necessarily posed a challenge to Government economic policy, towards the NCB, and towards energy and employment issues.

In the 1920s, the industry faced immediate and long-term prospects of decline. Although coal was to remain almost exclusively the source of primary energy for the next twenty years, the existing levels of employment and output were never to be reached again. Export markets were in the process of being permanently lost. By the 1980s, however, coal has already faced a number of competing primary sources of energy – oil, gas, nuclear power, and, in a sense, conservation. And it has already experienced a massive period of decline so that its economic prospects in Britain and the world more generally are regarded as highly favourable (by all but the Government and the dwindling nuclear power lobby).

The economic context of the two disputes also reveals differences. The immediate justification for the attack on the miners in the period leading up to the General Strike was the effect of the return to gold on the export competitiveness of the industry. The policy of returning to gold reflected financial interests but these coincided with and consolidated a weakness in the forces behind the alternative to direct attacks on labour, the rationalization of industry. To this extent, within industry, the interests of smaller-scale capitalists prevailed over those of large-scale capital. In the 1980s, finance remains a crucial determinant of Britain's economic policy and fortunes, albeit with a changed national and international role, but within industry the economy has become dominated by multinational corporations. These have relied upon and reproduced a low wage, low investment, low productivity economy in the UK, and Government policy has supported and consolidated the associated pattern of decline. Within particular sectors, such as energy, these corporations have been able to press for policy in their own interests and these have been

The coal question

disadvantageous to coal. Consequently, economic and energy policy has held back the coal industry, despite its favourable long-term prospects, with nuclear power particularly being pursued even though it is expensive and unproven.

In the interwar period, the coal industry had just embarked on mechanization in which coal cutting was important as well as the rationalization of extraction into larger, well-organized units. Britain lagged behind in these processes when compared with its rivals as a result of the economic organization of the industry. Where mechanization did proceed, it created a pool of unemployed miners who would be available at lower wages in backward pits, thereby maintaining a certain level of employment and commercial viability in these pits. By contrast British Coal has developed the most advanced techniques in the world for computer-monitored, heavy-machinery, deep mining. It is heavily investing in such methods in the central coalfields. When set against the low levels of investment in the peripheral coalfields and the overall target levels of output, the result is to create pressure for closures and redundancies. In short, whilst the private and fragmented industry before the war was ill-placed to face falling demand and the challenge of rationalization and mechanization, the current prospect is of a capital-intensive industry in public ownership going through further reorganization in preparation for privatization.

Taken together, these different economic factors have resulted in a significantly different political aspect between the two strikes. In the 1920s, the representatives of liberal, reforming, rationalizing capital could pretend to offer a resolution to the conflicts between miner and mine owner and between employee and employer more generally. Co-operation between capital and labour in reorganizing industry could lead to higher productivity to the mutual benefit of both parties. Of course, in the short term, labour would have to make sacrifices in order to maintain the commercial viability of existing enterprises – until the fruits of higher productivity became available. Such a position emerged explicitly in the General Strike and was to be identified with Samuel. It was to give rise to the Mond–Turner talks later in the decade.

The situation in the 1980s has had no such parallels. Keynesianism as the conventional economic opposition to monetarism, has no particular strategy to offer for the future of the coal industry. The Labour leadership is relinquishing its commitment to public ownership and extensive long-term co-ordination and planning of industry. This leaves an intellectual vacuum in the field of alternative economic policy making. It is one that the Tories intend to occupy through privatization.

Appendix 9.1: Evidence as presented to the review of Cadeby colliery

The economic costs of closure

Throughout, it is assumed that closure of Cadeby will result in all miners being made redundant in the immediate future over which the calculations are made (five years). Glyn (1984: 8), citing NCB figures, has noted that only 0.2 per cent of 43,141 redundant miners have been re-employed. In the Dearne Valley area, unemployment is 50 per cent higher than the UK average and stands at 1 in 4 for males. For Conisborough, the overall unemployment rate is 27.2 per cent and almost 1 in 3 males are unemployed.

Although some miners may be transferred within the industry, this is only liable to be at the expense of other miners (taking early retirement, for example, and the costs of this will be greater the older the 'substitute' redundancy) or of job opportunities for the miners of the future who would otherwise be able to enter the industry. It is also assumed that where redundant miners do find employment in other sectors of the economy, this is at the expense of others in the labour market who will be forced to remain unemployed in the miner's place.

In short, a job is lost through closure and does not, in present economic conditions, create an alternative job to take its place. The transfer of unemployment within the industry or to another sector or location will do little other than to affect the cost of unemployment slightly and who is to receive the burden of that cost.

The direct costs of closure

(a) Costs to British Coal

Under the Employment Protection (Consolidation) Act 1978, British Coal is required to make statutory redundancy payments. These are at the rate of one week's pay for each year of service. Currently at the colliery, there are 3 workers over 50 years old and 321 under 50. We shall assume that the former each have 35 years of service and the latter an average of 15 years' service. At an average wage of £165 per week the statutory redundancy payment would be $(3 \times 35 \times 165) + (321 \times 15 \times 165)$, £811,800. Of this, British Coal is liable for 59 per cent or £478,962.

(b) Costs to central government

Central government contributes 41 per cent of statutory redundancy pay which totals £332,838. In addition, the Redundant Mineworkers Payments Scheme, which is administered by British Coal but funded by central government and the EEC, involves a number of elements. One is a further £1000 lump sum compensation to those workers under 50 for each year of service, a total of 321 × 15 × 1000, £4,815,000. Assume a lump sum payment to each of the three workers over 50 of £14,280 (a total of £42,840). These workers also each receive an Unemployment Benefit Equivalent of, say £2,000 p.a. (a total of £6,000 p.a.) and a Pension Supplement of, say £500 p.a. (a total of £1,500 p.a.). The first of these two payments, which begins in the second year of redundancy, is half financed by the EEC with its paying an equivalent first-year contribution set against lump sum compensation.

As a result of the RMPS, central government will incur a lump sum cost of £4,862,340 and a recurring cost of £4,500 p.a. The EEC will incur a recurring cost of £3000 p.a.

Central government will also have to pay unemployment benefit in the first year of redundancy at a weekly rate of £30.45 for a single miner and £49.25 for a miner plus dependent spouse. Assume there are equal numbers of single and such married miners. Then total payment made is £665,176.

Ninety per cent of the cost of rent and rate rebates are paid for by central government. In the Dearne Valley, only 40 per cent of households are owner–occupiers (56 per cent in the UK as a whole) and it is only 1 in 3 in Conisborough. Assume that 160 redundant miners receive a 50 per cent rebate on a total rent and rate payment of £1,000 p.a., then recurrent cost to central government is £80,000.

In each year, a miner in employment would be expected to make an income tax and national insurance contribution to central government of approximately £1,800. These would be lost if the miner were made unemployed together with indirect taxes, such as VAT, on consumption (which would be lower with loss of income). Leaving the latter aside, the total loss of central government revenue would be £577,800 p.a. for 321 workers.

(c) Costs to local government

Local government would lose the rates paid by British Coal which are £86,340 in 1985/6. In addition, it has been estimated by Sinfield and Fraser (1983) that local government spends approximately £500 for every unemployed individual for the provision of the resulting extra services provided. This would add £160,500 p.a. to local government expenditure for 321 workers. A further cost is 10 per cent of rent and rate rebates, totalling £8,000 p.a.

All the above costs are summarized in table 9.1.1.

Table 9.1.1. Total direct costs of closing Cadeby colliery (£000)

	First year after closure	Annual costs	Sum of costs over 5-year period
British Coal			
Redundancy payment	479	—	479
Central government			
Redundancy payment	333	—	333
RMPS	4862	5	4880
Unemployment benefit	665	—	665
Rent and rate rebates	80	80	400
Income tax and NI	578	578	2890
Local government			
British Coal rates	86	86	432
Rent and rate rebates	8	8	40
Extra expenditure	161	161	805
EEC			
RMPS	3	3	15

The indirect costs of closure

If Cadeby were closed, British Coal suppliers would lose orders and this would result in further job losses in the firms concerned and to the firms supplying these and so on. We assume an economic multiplier of 1.2, i.e. for every £ of lost NCB expenditure, a total of £1.20 is lost to the economy as a whole. For Cadeby, non-wage costs are approximately 50 per cent of operating costs and might be expected to be £10 million in a normal year. Hence, £2 expenditure might be lost to the economy over and above British Coal's own expenditure. If we assume this were spent on other industries with a value-added of £10,000 p.a., then a loss of 200 jobs would be felt by the economy. These lost jobs would also result in extra central and local government costs for unemployment benefits, etc. These would result in costs of approximately £2.6 million and £300,000, respectively.

By a similar logic, there will be a loss of jobs because of the reduced expenditure by miners as a result of lower income. Assume a multiplier of 1 on an annual net decrease of expenditure per miner of £5,000 per annum, then the economy would lose £1.5 million or 150 further jobs and central and local government would incur costs of £2 million and £200,000, respectively.

Appendix 9.2: The economic case for coal

This is a slightly amended version of evidence presented to the review of Cadeby Colliery. It depends heavily for its content on Kerevan and Saville (1985b).

The coal question

The economics

The NCB's decision to close Cadeby, subject to review, is part of a wider strategy to close so-called 'uneconomic' pits. This strategy is to be implemented through an objective of only continuing capacity whose selling price will on average fall at around £40 per tonne. It is important to recognize, however, that this strategy is not unique to the NCB but that it runs along the same lines as proposals emanating from the EEC. Whilst it might be thought that the common thinking of the EEC lends weight to the strategy of the NCB, this is the opposite of the case since the assumptions on which the strategy is based are faulted and, the wider the strategy is adopted across the EEC, the more it is certain to fail as the consequences of false assumptions are felt across a wider range of application.

For the EEC, coal capacity is divided into three categories; fully competitive (50–60 million tonnes), on the verge of competitivity (140–150 million tonnes), and hopelessly uneconomic (40–50 million tonnes). This means that for break-even production, output would be at about 100–120 million tonnes or half recent Community output. The logic on which the EEC intends to push through a closure plan and withdrawal of subsidies is through reference to labour productivity levels and this is faulted by failing to take account of the cost of capital in the economics of different pits. This is of less concern to us here than the false logic by which half of EEC capacity is designed as uneconomic.

Essentially, this depends, not on a reduction of coal consumption in the EEC in the future (in fact considerable growth), but on the replacement of EEC production by imports. This is justified on the grounds that import prices of coal are currently cheaper than EEC production costs and that they will continue to be so in the future. The current prices of coal on world markets is severely depressed as a result of the coming to availability of capacity from investments of ten or more years ago, these being made in the wake of expected continuing (energy) growth and this then giving way to preference for coal after the oil crisis. Consequently, there are many fixed investments in capacity that have been made and whose costs cannot be recovered in conditions of excess supply. This leads companies to attempt to sell at any price that will at least lead to the covering of current operating costs even if higher prices can be obtained in more secure domestic markets. The result is low prices on world markets which do not cover long-run operating costs.

This leads to three fundamental conclusions:

1. By designating EEC coal production as uneconomic because it has overall costs which are higher than operating costs of imported coal, a double standard is being applied at the expense of EEC coal. Whilst importers only have to cover operating costs, EEC producers have to

Privatization and property rights

cover full costs. This is not only, in some sense, unfair, it is also poor economics. This is because the capital costs of EEC pits are already sunk and are not retrieved by closure and this explains why importers are prepared to sell at below full costs. As is recognized in any industry during recession, survival depends on covering operating costs and as much of fixed capital costs as possible but these latter may have to be written off. Such a process is more acute in the coal industry because of the long gestation of fixed capital expenditure before there are revenue returns. As already mentioned, the EEC strategy depends on relying upon the importers to pursue this strategy but on preventing EEC producers from doing so.

2. By displacing EEC coal by imports (to cover expanded demand into the future) the nature of world trade will be transformed with EEC imports rising from 60 million to somewhere near 200 million tonnes, almost doubling world trade at current levels. Even in conditions of excess supply, if they continued, it cannot be imagined that this change in the nature of world markets and the strengthening of the hand of what are liable to be relatively few suppliers will leave import prices as depressed as they have been as producers attempt to recoup more and more of fixed costs in export rather than domestic markets. Equally, current low prices of delivered imported coal depend crucially on the severely depressed state of the shipping industry and this would be modified by the expansion of coal trade. Moreover, it cannot be presumed that expected sources of imports are necessarily secure, nor even desirable, with South Africa being seen as a major exporter.

3. Perhaps, most important, the current depressed levels of coal (and shipping) prices are not the basis on which to formulate policy for the future. As and when excess supply disappears from world markets, import prices will have to cover full costs including capital programmes. In this light, world market prices will be much higher than those at present and above what are anticipated to be the costs of production of much, even most, of the supposed EEC uneconomic capacity. It is perhaps worth observing that the CEGB's own case for Sizewell would not stand if the same price criterion for mine closures were applied (as imported coal-fired stations would prove cheaper).

The ideology of subsidy

Whilst the previous section details the false economics on which the EEC coal policy is being based, there is no doubt that this false economics is fuelled by the idea that the industry is in receipt of massive subsidy which will vanish with a closure programme. This is the opposite of the case since much of the current 'subsidy' will continue regardless of closure – such as redundancy and interest payments from the past, for

The coal question

example, and some of these costs will be increased, not saved, by closure. Details of these factors for Cadeby are calculated separately (as in Appendix 9.1). Similarly, the costs of closure incurred by the economy, if not by coal companies, include loss of taxation and the need to make unemployment payments. Again, these are calculated separately for Cadeby.

Macroeconomic effects

By relying on a high import strategy for coal, the EEC will have to spend much foreign exchange (probably dollars as this is the currency in which the world market is quoted). This will have adverse effects on balance of payments and, together with increased budget deficits due to the costs of closures and the loss of employment income in mining and dependent industries, macroeconomic performance and the room for manoeuvre for macroeconomic policy will be seriously impaired.

Chapter ten

Coal: the ultimate privatization

Earlier chapters have criticized the new approaches to the problems posed by privatization. In the first section, an alternative analytical framework is offered, drawing upon that proposed by Fine and O'Donnell (1985),[1] whilst offering comparisons with other approaches, particularly with the synthesis. This is followed by an application and illustration of the approach by reference to recent developments in the coal industry. Finally, there is a brief assessment of the emerging schemes for privatizing the industry even though these are necessarily speculative at the time of writing.

Towards an alternative theory of privatization

Any account of privatization must employ an economic theory, however weak or implicit. The one adopted here is associated with many of the insights of Marx's analysis of capitalism, although much would also be accepted within other unorthodox intellectual traditions, such as institutional economics and post-Keynesianism. In particular, emphasis is placed upon the tendency for capitalism to increase the capital intensity of production, the concentration of production and ownership, and for these to be associated with wider economic and social changes which prove to be sources of conflict between capital and labour.

In the modern period such economic processes have created pressures for state regulation, not only to provide a counterweight to the potential abuse of monopoly power, but also to facilitate the changes involved – in the interest of competitiveness – at the level of productivity as well as in the market place. This, in turn, has given rise to social as well as economic criteria for the running of industry, to the point of public ownership, and even by excluding certain activities from immediate commercial criteria altogether, as in much welfare provision.

Thus, any consideration of privatization, as of public ownership, must be fully grounded in an economic analysis but also in the relations between economy and society – in order that the extent, nature, and significance of the state's interventions can be comprehended. Otherwise, the role of the state's interventions will be reduced exclusively to those of a peculiarly acting private agency – even if its motives and constraints are determined externally to the theory. In the framework of analysis suggested here, as in the synthesis, there are (different) underlying economic forces at work, but these are now inseparably linked to broader social influences that are able to accommodate a role for class and other conflicts at the economic and political levels. Thus, whilst the proposition of no general theory of nationalized industries is accepted, as in Vickers and Yarrow (1988), in the light of the complexity of the influences at work, there is an emphasis on a broader set of factors than those of the synthesis and with a different understanding of those that are considered in common.

This leads to an alternative understanding of the history of the nationalized industries. Prior to being taken into public ownership, the industries concerned had experienced failure in developing along the lines indicated above. Far from moving into a condition of natural or artificial monopoly, it was more often a case of a fragmented industrial structure unable to reap economies of scale. More generally, it can be argued that capitalist relations of production had been eroded, not simply in terms of lack of profitability but across the range of activities and conditions that are associated with commercial enterprise.[2] Whilst wage labour and the sale of products as commodities continued, price levels were heavily regulated, technology had become backward, industrial relations were strained and, following the post-war Labour victory, political conditions were far from favourable to the continuance of private ownership in the industries concerned. To some extent, these developments were a product of wartime controls, but they were also the result of the industries' experiences in the interwar period.

The subsequent history of the nationalized industries can be seen as a process of restoring them to commercial viability in the broad sense of re-establishing the relations of management, technology, finance, and marketing to conditions in tune with if not identical to those associated with the private sector. In this light, the current policies of privatization are not so much a break as a continuity with those of the past. It becomes relevant to question why those industries that have been privatized could so easily be returned to the private sector, even if not on the terms and conditions desired by the synthesis. It reflects more than a contingent movement in technology, etc., prejudicing the presence of natural monopoly, for the ability to privatize represents the culmination of the process of reorganizing the industries concerned, a process often beginning with the first days of public ownership, in the light of the management appointed and the structure of management to exclude workers' control/participation. Even prior to the current Government, the

Labour Government's preoccupation with commercial criteria led to a White Paper on the nationalized industries in 1978 which was simply taken over by the new Tory regime and implemented through cash limits (Fine and O'Donnell 1985).

Nor is the current phase of privatization to be entirely explained by a judicious combination of the political ideology of the Thatcher Government, with its supposed commitment to the free play of market forces, and the wish to fund short-term tax cuts from privatization revenues – important influences though these may be. Apart from other political motives such as 'restoring the balance in industrial relations', a major factor behind privatization is to be found in developments associated with new technology. These have had the effect of breaking down the traditional boundaries that divide one sector of the economy from another. New technology has been generally applicable, adaptable, and cheap, with potential use in production, management, and marketing.

Accordingly, the economy is going through a particularly intense period of reorganization along the lines of vertical and horizontal integration, especially where new technology is concerned. And, once this has been embarked upon, it even incorporates sectors that might not be going through substantial technical change, or be related directly to those that are. For merger and acquisition between firms that are already diversified throw up a more general process of restructuring across sectors as corporate production portfolios are rationalized.

The prime example is given by the newly created sector of information technology. Previously the sector of office equipment and data processing and of telecommunications were essentially distinct. Now they are intimately related to each other so that companies confined to one or other of the sectors are obliged to engage in both sectors. In the case of British Telecom, this has been made possible (for it to move beyond telecommunications) by privatization whilst its new competitors have been allowed greater scope through their liberalized access to the telephone cable network. What is certain is that the pre-existing juxtaposition of private and public sector firms could not have persisted following the advent of the new technology, despite the scope for joint ventures, although privatization of British Telecom was by no means an essential outcome.

A further impetus to privatization has been provided by the general economic conditions. Following the crisis of 1974, there have been extremely high levels of unemployment and this has been compounded by a number of other factors. Deindustrialization has entailed a particularly severe rundown in the, predominantly male, manufacturing workforce; for demographic as well as for other reasons there has been, especially in the UK, a growth in the female labour force engaged in part-time, low-pay service sector employment; and the Government, since

1979, consciously or not, has generated high levels of unemployment through constraints on the growth of aggregate demand. All of these have acted, together with other measures such as reform of welfare and employment and industrial relations legislation, to worsen the bargaining position of the workforce as a whole and those within the nationalized industries. In addition, private capital has been that much shorter of avenues in which to employ its capacity and finance and this makes the destruction of the places for public sector 'incumbents' that much more attractive.

Finally, there is an alternative to the synthesis at the level of policy which takes as its starting point the longer-term problems of the British economy and an assessment of them. These are best described as the 'three lows'. Within the advanced capitalist countries, Britain has become the low-investment, low-productivity, and low-wage economy. Indeed, British manufacturing has only remained internationally competitive to the extent that it meets greater productivity increases elsewhere by cutting wages relatively. This is, however, only to describe Britain's economic problems cursorily, it does not explain their causes.

Clearly, there are many explanations for Britain's economic decline, with its vicious circle of low wages and productivity, and we cannot review them here. Elsewhere, Fine and Harris (1985) have argued that it is a consequence of the lack of any economic agency committed to and powerful enough to adopt a coherent and long-term strategy for the regeneration of British industry, both sector by sector and by co-ordination across sector. Whilst British financial institutions have been more concerned with international operations, more often than not to the relative neglect of long-term commitment to and supervision of industry, British industry has itself been the most dominated by transnationals relative to its size (and second overall in absolute terms next to the United States). Accordingly, industrial 'policy' has been more responsive to the global planning of transnationals and this has helped to draw upon and reproduce the UK as a source for a 'three lows' labour force, often serving as the assembler for the rest of western Europe. Moreover, the labour force itself, whatever its role in claiming too high wages and obstructing changes in work practices, has had limited effectivity, whether through the trade unions or the Labour Party in Government, in creating a long-term industrial policy.

This has meant that government policy for industry has lacked direction, even if it has been far from minimal. Instead, however, it has been subject to a complex of competing interests in which overall co-ordination has been lacking. No doubt continuing commitment to and even extension of public ownership would be no guarantee against this persisting into the future, as in the commitment to nuclear power in the past. But the case can be made that it is a necessary condition for the reversal of Britain's deepening economic problems. As an alternative, the synthesis offers not so much a blind faith in the market, as the *laissez faire* privateers would propose, but more the equivalent of a lapsed Catholic, recognizing both the virtues of the market and the (sinful) benefits of regulation. With the Jesuits in power, their influence is liable to be limited both in its own scope and in its acceptability.

From tonnes to money

There has been a traditional view, lasting a century or more, that the future of the British coal industry is primarily determined by the unyielding logic of deteriorating geological conditions. Productivity is inevitably bound to decline, with the industry itself, as the most accessible reserves are extracted. As discussed in chapter 3, the most famous economist to put forward such a pessimistic outlook was W.S. Jevons, in the mid-nineteenth century (Jevons 1865).[3] David Ricardo had previously popularized a similar argument for agriculture, with the decline of the British economy being laid out into the future as decreasingly fertile land was brought into cultivation to feed a population growing at a geometric rate in Malthusian fashion. Jevons took the argument one stage further. Through exporting coal, Britain could pay for food imports. By this means, economic decline could be delayed, if not prevented, since the exhaustion of coal seams underground merely displaced the infertility of farm land above ground – eventually decreasing productivity of British coal would confront decreasing Ricardian productivity of overseas agriculture.

Jevons predicted that this sorry state of affairs would have come about in the 1950s or 1960s, just as the declining use of coal was about to begin. But his theory is clearly empirically fallacious, as is the Ricardian theory of diminishing returns in agriculture, so rapid has productivity increase been. Productivity in coal-mines has continued to rise and estimates of new or recoverable reserves tends to outpace extraction. Also, alternative fuels, notably oil and natural gas, would appear to have had more to do with the decline of coal than deteriorating geology.

Nonetheless, old ideas die hard. Supple (1986) argues that the twentieth-century coal industry is caught between declining productivity and declining demand, pincered as it were between supply and demand. Consequently, its economic and social history involves painful adjustments to such underlying material realities and, inevitably, economic, political, and social conflict between employers, on one side, and miners, their communities, and their political supporters, on the other, with the government (when not employer) caught in the middle. When the state is the employer, this may lead to it serving as the focus of conflict, and an unduly moderate pace of decline.

Whilst this is a potentially neutral stance over the extent and pace of change (equals decline) and the more or less favourable conditions for miners under which it occurs, the logic is one of fiercely resisted decline with repercussions over and above the narrow economics involved. There is also the presumption that state control and ownership of the coal industry cushions decline and leads to governments seeking a quiet life granting, as it were, the status of 'estate' to the miners and their communities. This has, however, its counterpart in the idea that privatization will sweep aside the privileged position of miners and in a short, sharp shock restore the industry's decline to its rightful place and pace, alongside in the British case, with the import of cheap coal and the development of nuclear power.

This whole approach has been severely questioned by O'Donnell (1988). She places much less emphasis on the deteriorating configuration of supply and demand, with the latter presumed more prominent in the modern period. Instead, by comparing the pit closure programme of the 1960s and of the 1980s, she points to the crucial role of government policy in determining the size and performance of the industry. In practice, and oversimplifying, demand has been set by a target output and the conditions of supply by the allocation of investment funds. These are both a matter of policy and there is no presumption that a higher output target means higher (marginal) costs because of declining geology. For the extra production could have been provided at much the same cost as continuing production, if sufficient investment funds had been made available to match capital equipment in use.

Indeed, what has occurred, and this explains much of the so-called tail of uneconomic pits, is that output targets can only have been reached by production on pits in which there has been underinvestment. Significantly, in the first decade of nationalization, domestic prices were held below world levels, coal imports were subsidized by the NCB, and investment plans for long-life pits were not and could not be achieved because of the limitations of internally generated funds and on externally provided governmental finance. Consequently, the closure programme of the 1960s was not simply a response to cheap oil but reflected the lack of capital invested in what were to become uneconomic pits.

Nor is the history of the British coal industry under nationalization simply one of cumulative disadvantage through government policy – although this does not imply that the fragmented private industry would have been preferable. For, as Fine *et al.* (1985b) have argued, the previously eroded conditions of capital's control of the industry, which made nationalization essential, have been increasingly re-imposed. For the industry has been qualitatively transformed – in marketing, investment criteria, management, and industrial relations, technologically and in the production process. These are now examined further in the light of the more recent developments towards privatization.

First, the overall criteria for operating the industry have been changed. Previously, the main instruments were the adoption of a target for output and some form of cash limit, usually unachievable. According to the Monopolies and Mergers Commission (1989: 17), British Coal's own view has been to 'turn an institution into a business' and the Commision itself points to a change in culture from 'tonnes to money' (p. 5). This is closely linked to investment criteria which take as their target the level of world coal prices.

Accordingly, the overall benefits of investment to the economy are eschewed in return for the requirement of marginal cost for new investment of £1.00 per gigajoule (GJ) for new output with overall operating costs in the colliery concerned to be below £1.50 per GJ (see NCB 1985). This is despite anticipated revenue of approximately £1.65 per GJ. But the exact figures are not so important as the narrow basis on which the calculations are made as far as the benefits to the economy are concerned. On the other hand, the imposition of such narrowly commercial criteria are conducive, if met, to the restoration of the industry to the private sector.

Second, the pricing policy of British Coal has been heavily determined by its relations with the CEGB – which merely has the effect of transferring surplus or deficit between one nationalized industry and another. The guideline increasingly used, however, has been the seriously depressed levels of world coal prices. It has had the result of the South of Scotland Electricity Board and British Coal conducting an eighteen-month long running battle over coal imports, with the latter continuously holding the threat of court action over the former. Meanwhile coal imports have risen even though this is of net disadvantage to the economy as a whole.

Third, there has been a transformation in the management objectives of British Coal. This has been clearly stated by Mr Northard, then Operations Director now Deputy Chairman of British Coal, in evidence in 1987 to the review of the Manvers Colliery Complex:

> British Coal is not a social service, and no amount of wishful thinking can change the situation, – only the Government can decide how we should operate as an industry.

This clearly, stretching the meaning of 'social service', is a total breach with the aims of nationalization, to run the industry to break even taking one year with another and to take due account of the welfare of the workforce. The Energy Select Committee (1986/87b: li) see this as a move towards commercial operation in response to the 1978 White Paper on the nationalized industries, although this is 'imprecise':

The coal question

The 1978 White Paper on the nationalised industries essentially required the industries to act as commercial entities and therefore, presumably, to ignore wider social and economic costs and benefits.

Consequently, forcing a division between commercial and wider economic and social considerations, the Committee concurs with the Government's conclusion that:

> the task of dealing with the consequences of coal policy to mining communities rests in central government with the Departments of Health and Social Security, Employment, Trade and Industry and Environment, together with the Scottish and Welsh Offices.

With so many minders, one might have expected the miners and their communities to be secure. However, whilst it is claimed that this is indeed done (see page 139), the Government refused the Select Committee's suggestion that this should be ensured through a special Cabinet Committee to co-ordinate the separate agencies involved (Energy Select Committee 1986/87: 5). In short, this is a perfect example of decision by compensation criteria with the compensation not being made! (See chapter 8.)

Fourth, the prospects for privatization have been enhanced by the weakening and splits within the workforce. These do not simply derive from the formation of the breakaway UDM in the course of the strike of 1984/5, but they have much deeper roots. As Fine *et al.* (1985a) argue, the unity and militancy of the union in the early 1970s was a product of very special circumstances, including national measured day wage bargaining to the advantage of those at the bottom catching those at the top of the pay league within the industry over both grades and regions. These conditions conducive to militancy were eroded by the negotiation of district and other productivity deals from 1977 onwards. With higher wages and investment in the central coalfields, divisions between the area workforces were always likely, weakening the union as a whole and preparing the way for potential privatization of different sorts of pits separately – whether high wage with high productivity and high investment or, alternatively, with all three of these as relatively low.[4]

Fifth, British Coal has been introducing the most advanced underground mining methods involving electronic monitoring, heavy duty machinery, and the more effective retreat mining on fewer faces in fewer pits.[5] This is ultimately the method by which privatization can be secured for, with greater pace and intensity of work through increased machine time available and with six-day working, individual pits will be guaranteed profitability with appropriate capital write-offs for the investments that have been made. This is the basis on which so many

Coal: the ultimate privatization

jobs have been lost in the industry with relatively small decline in output, and it equally points to the fallacy both of the necessity of uneconomic pits and the inevitability of decline through geological imperatives.

Finally, in analysing the transition towards privatization, it is as well to be aware of what has already been achieved. Whitfield (1985) has discussed these possibilities, revealing the extent to which the industry is already penetrated by the interests of private capital – in opencast, machinery manufacture, by-products, and land sales, for example. There have also been the selling off of subsidiaries and the subcontracting of services and even mining operations such as the driving of tunnels, often by those who have taken redundancy. This reveals the extent to which the idea of the nationalized industries as a monopoly restricting the entry of private capital is a partial view, for the nationalized industries have been confined to their sectors of operation rather than being allowed to diversify into areas where their existing activities lend them an advantage. But despite the existing penetration of private capital in and around British Coal operations, there are particularly acute difficulties to be faced by a strategy of privatization. Recognizing these reveals the form that any privatization might take. But, as in the case of electricity, the existence of such problems is not likely to lead to a reconsideration of the overall costs and benefits.

Concluding remarks

In late November, 1986, the Prime Minister confirmed that, 'there are no current plans for the privatisation of BC'; in early 1987, in response to the Energy Select Committee (1986/87a: lxxiv) the Energy Minister, much to its disappointment, 'chose to caricature the call for greater private sector involvement in the British coal industry as "absurd" and even "lunatic".' Within eighteen months, and following a general election, the new Minister, Cecil Parkinson announced that British Coal would be privatized. Presumably, the government felt that its condition for privatization, announced in May 1987, was about to be achieved (Energy Select Committee 1986/87: 7):

> By the end of this parliamentary session the Government will have privatised two fifths of the public sector industries which it inherited in 1979. It has also announced its intention to return most of the remainder to the private sector in the next parliament. There can therefore be no doubt of the Government's commitment to privatisation where this is appropriate. But the priority for British Coal is to restore the industry to financial viability. This is why there are no plans at present to privatise British Coal.

The methods for achieving financial viability have been discussed in the previous section. In addition, the Government appears to be motivated by a two-fold strategy of preference for nuclear power and hostility to the miners, all bound up in an ideological preference for market forces which specifically involves coal imports in the short run, whatever the long-run and wider economic and social implications.

It might be thought that the policy of greater and freer imports would reduce the appeal to investors of the British coal industry. This is, however, to view the issue too abstractly. For those who are liable to be involved in purchasing British Coal are the large-scale, usually multinational, companies already involved in energy and often coalmining. They favour both privatization and freedom to import, together with the building of port facilities to enable greater imports. What they seek is both UK markets for their coal produced abroad, and a share in that part of the British coal industry that will guarantee profits through high levels of productivity and sufficient capital write-offs (or low enough selling price).

Illustrating this are a number of schemes that have emerged in discussing privatization. British Coal has itself been investigating diversification into electricity generations. It is studying a joint venture for a new power station, privately funded, and requiring no-strike agreements from the UDM. Esso is reported to be interested in bidding for the North-East coalfield.

So is Consolidated Gold Fields, a British-based company with South African links in gold mining, and interests around the world in mining and quarrying. It has a joint venture with Ryan International, one of the largest private coal producers who are currently licensed by British Coal. Whilst the private producers are extremely small, in the context of privatization they will have a strategic role in view of their incumbent presence.[6] Perhaps more important, the managing director of Consgold is Allen Sykes, a leading advocate of privatization of the coal industry and co-author of a pamphlet, published by the Centre for Policy Studies, arguing for privatization, on which see later.

In 1988/9, Consgold has been under threat of takeover by Minorco in what would be the largest ever bid, £3.2 billion, on the UK stock exchange. Minorco is the foreign investment wing of the South African-based company Anglo-American. This is a huge multinational, owning two-thirds of all shares quoted on the Johannesburg Stock Exchange and the largest single investor in the United States.[7] By taking over Consgold, and it has been prepared to divest many of its interests in gold, platinum, and rare metals to satisfy monopoly investigations, Anglo/Minorco would establish a presence for itself in the British coal industry both as producer and as importer. This is crucial for a company that needs to diversify in general from South Africa as the crisis of apartheid deepens; and it is crucial in the particular need of finding a way around increasingly stringent sanctions against South African coal.

Such commentary is necessarily speculative at this stage but it serves to illustrate the extent to which privatization of the coal industry will lead to the making of economic and political policy outside the direct control of government. Indeed, there is a sense in which exactly the opposite is true. The Government wishes to adopt a policy of importing (South African) coal whilst presenting this as a market rather than as a political decision.

There are still, however, problems with privatizing coal. First, the NCB is a highly integrated and centrally co-ordinated operation, particularly when it comes to the introduction of new technology and the allocation of markets. In the case of new technology, this has to be tried and tested in place. The experience and knowledge of the machinery as well as of geological conditions and how to cope with them is gathered across the whole of the industry. As Winterton (1985: 233) observes of the new technology Minos facility:

> Before the strike about eighty pits had at least one MINOS facility, but none had the entire system; the first complete MINOS installation was to have been the Woolley complex in Barnsley.

Consequently, privatization of particular pits would face the problem of what would be the continuing relation between British Coal and one of its competitors when British Coal has the expertise in new technology.

Other than selling off British Coal *in toto*, which would prove difficult given its paper losses, one solution might be for a multinational energy utility, with some experience of deep mining in other countries, working in co-operation with a revamped British Coal in running privatized pits. This is not impossible, for similar developments have already occurred in telecommunications, where British Telecom has been obliged to co-operate with Mercury in setting up the latter's system, and in gas, where the British Gas Corporation has been forced to make its pipeline available to North Sea oil producers. But there are still further political and ideological problems in so far as British Coal will be supporting not only a competitor but also one that is a multinational corporation with limited previous connection with the UK.

There appear to be two alternatives. In one, the R&D and other overhead skills of British Coal are pooled as a central service to all producers. In the other such central facilities are abandoned altogether and they are devolved on individual producers. This latter seems the most likely outcome as, since the strike of 1984/5, British Coal has been running down its overhead staff and concentrating greater autonomy into fewer, larger areas.

Not surprisingly then, the regional areas are the preferred units for privatization. One of the earliest contributions in the demand for privatization, Boyfield (1985), argued for the creation of separate area companies prior to privatization. More recently, Robinson and Sykes (1987) have also stated a preference for such a structure, rejecting both a single company and the privatization of pits individually.[8] They calculate that privatization will be of enormous benefit – in the range of £460 million to £635 million annually from 1989 to 1992 and in the region of £950 million plus and rising, subsequently.

Their estimates are entirely spurious for a number of reasons, even where they are transparent. First, they translate a reduction in the price of coal into a benefit even where it simply reflects a transfer price (so the cost will be felt by somebody else). Second, the benefits are simply asserted in many cases to flow from privatization through increased productivity and management efficiency, and these impose costs on the workforce. Third, the positive value of British Coal's assets as realized in a sale are counted as a benefit, even though they are there irrespective of privatization. Fourth, the only sensible case that can be made in their terms is through cheaper imports of coal and closure of uneconomic pits. As has been argued previously, however, such benefits are outweighed once account is taken of long-run and wider economic and social benefits.

Much of the above concerns only the privatization of the most advanced pits, but so-called uneconomic pits could also be targeted for privatization. These could be made profitable, in the absence of substantial investment, only by substantial worsening of the wages and conditions of the workforce. Here, as in the 1930s, the miners released into unemployment by the high technology mining will provide a pool of unemployed which is potentially available at low wages for low technology production. But this is precisely what has lain behind one component of the Government's privatization strategy – to allow private capital to exploit as fully as possible the weak position of workers in the labour market that has arisen from high levels of unemployment.

In either case, there still remains the problem of marketing. In continuing conditions of excess supply and with the CEGB the major customer for coal, there is the issue of how markets will be allocated to producers – who gets the contracts? There is the further complication that coal comes in a number of different grades with a variety of different properties so that mixes are used to meet but not to exceed customer requirements for quality. Again co-operation will be required in supplying, pricing, and allocation of revenues, something which has been a feature of British Coal accounts and their effect on measured colliery performance. Only, post-privatization, there is the possibility of marketing being heavily influenced by corporate vertical integration into electricity supply.

Finally consider the question of royalties. This has been much discussed again after an interval of fifty years, partly because of the wish to liberalize the conditions and lower the costs in which the private mines operate and, much more importantly, because control of the royalties is control over who mines and where. Much of the discussion is ill informed both historically, being unaware of the price that British Coal paid for the royalties in the past, and practically. Prior to the decision to privatize, the Department of Energy had laid out the problems involved in response to the Energy Select Committee's proposal that royalties should be vested in the Crown with the Government issuing licences to mine, presumably through the Department of Energy (Energy Select Committee 1986/87b: 7-8):

> 170,000 mining reports are issued each year to solicitors and others, many of them acting for potential house purchasers . . . making good (or compensating for) subsidence damage caused by lawful working of coal, whether or not the mining was carried out by British Coal . . . fencing off abandoned and disused mines and quarries . . . . There are well over 100,000 abandoned mineshafts in all coalfields, some in quite remote areas . . . provision of mines rescue services and miners' welfare organisations . . . the sponsorship of research and representation of UK coal mining interests on international committees . . . . The responsibilities of British Coal associated with ownership of the coal reserves of Great Britain are wide-ranging and substantial. Any department or agency which took them over would need to employ several hundred extra staff and to have a large annual budget, possibly of £20-30 million a year. Its staff would have to include specialists in fields such as mine surveying, mining engineering, geology, civil engineering and mining law. Moreover, it would not be a matter of simply transferring staff from British Coal, since the Corporation would continue to need the services of their own staff in these fields to represent their interests to the new agency. The creation of this additional layer of bureaucracy would have to be set against the potential benefits.

This is a classic argument for public ownership in the presence of economies of scale. But more is involved. For whoever licenses the privatized industry will have a substantial degree of command over it, especially in the absence of the mining companies owning the land that they are going to mine. In the British context, this will create enormous difficulties for what will essentially be a body which will not only have to undertake the functions previously listed but also to act as a regulator of an 'input price' (royalty) and wide-ranging conditions of operation. It is worth recalling that the previous taking of the royalties into public ownership in 1938 was motivated by the need to reorganize and rationalize the industry. It appears inescapable that a new licensing authority would also have to take a view on and direct the development of the industry in some detail – in which case, it is far from clear to what extent the industry can be privatized in any meaningful sense as far as the supposed introduction of market forces is concerned.

This is not to doubt the Government's determination to see privatization accomplished even at the expense of storing up problems and inefficiencies into the future. Nor is there any doubt of the interest of private capital in making gains out of the return of the industry to the private sector. The results for the British economy, however, will be merely to consolidate the structures and mechanisms of relative decline.

Notes

Chapter 1 Monopoly capitalism and the coal vend

1. This chapter is based on Fine (1988a).
2. See Foster (1986) for what would appear to be an approved history and spirited defence against the many critics of the theory of monopoly capitalism as developed through the Monthly Review school.
3. See Sweezy (1942b) for discussion of Burnham and the managerial revolution and Sweezy (1943) for discussion of Schumpeter's theory of innovation.
4. For the details of the North East coal industry, see, for example, Nef (1932), Flinn (1984), Smith (1961), and the forthcoming Coal Board commissioned history of the period to 1700 by J. Hatcher.
5. There is a parallel with the nineteenth century US oil industry for which Rockefeller gained control for Standard oil through the distribution system.
6. Hausmann's models take no account, however, of the effect that the Vend may have had on cost conditions but simply measures the 'degree of monopoly' on the basis that these are given.
7. See also Cromar (1977) who makes a similar point about the eighteenth-century coal industry whilst observing the spatial expansion of the industry, as well as of its output, although his critique is primarily directed against Ashton and Sykes (1929). Unlike Hausmann, he does give some consideration to the cost structure of the industry, but not how it might have been affected by the Vend, other than as a potential barrier to entry. Nor is there, to anticipate a theme of later chapters and for a later period, any consideration of the role of landed property even though the Church is the dominant landowner in the area.
8. It is not clear that the primary and/or secondary sources on the relations of production are or could be adequate to establish how wage labour came about in the industry and how this affected and drew upon the economy and society more widely. Possibly, Hatcher's forthcoming history will be of assistance, particularly as he was a participant, from a non-Marxist point of view,

Notes

to the Brenner debate, Postan and Hatcher (1985).

Chapter 2 Cartels and rationalization in the 1930s

1. This chapter is based on joint research with Steve Martin.
2. Nor is this simple theoretical dichotomy satisfactory for it would apply equally well to whether traffic progresses at a roundabout or not when cars approach simultaneously. Does one (or all) forge ahead (with possibility of mutual destruction) or does one (or all) make way (with possibility of paralysis)!
3. This point employing the very same example has been independently made by Boyns (1987).

Chapter 3 Royalty or rent: what's in a name?

1. This chapter is based on Fine (1982a).
2. Church (1986: 768) has questioned the extent or even existence of productivity decline once account is taken of hours and shifts worked.
3. Ricardo argued that corn would become increasingly costly as agriculture moved on to increasingly less fertile land (the extensive margin) and this would eventually bring down profitability and cause capital accumulation to halt (the stationary state). Whilst Jevons rejected Ricardo's value theory based on labour time, he fully embraced his rent theory; see Fine (1982b).
4. *Reports of the Royal Commission on Mining Royalties* (1890–93), comprising five Reports, with Minutes of Evidence and Appendices.
5. The locational advantage of bringing 'coals from Newcastle' had always been their ease of transport to London by coastwise shipping.
6. For an analysis of land in Britain surveying many sources, see Catalano and Massey (1979). In 1926, the Samuel Report found that there were 3,789 royalty owners. Of these, the most fortunate 100 alone received over half of the royalty revenue. For more details, see chapter 4.
7. In part, the Commission believed that the industry's record spoke for itself and this alone proved that there was nothing wrong with the royalty system. Unfortunately, the same record of success did not exist for iron ore extraction, which suffered from international competition at this time. However, this particular empirical evidence was conveniently forgotten when drawing conclusions on mineral royalties as a whole. An exception to the Commission's findings was also to be found in Ireland's peculiar system of landed property, but this was also left aside.
8. For a more detailed discussion of the evolution and conundrums of

the neoclassical, marginalist revolution, particularly in relation to rent theory, see Fine (1982b).
9. Thus, there is J.A. Hobson's (1891) appropriately titled 'The Law of the Three Rents', used to explain wages, profits, and rents. The modern form in which the categories of political economy are conflated is around the concept of capital and rate of return to all factor inputs as assets rather than as a rent from each as comparable to land. This is most notable in 'human capital' theory in which rewards to labour are seen as a return to an educational asset.
10. This debate was surveyed by Buchanan (1929). He attempted to confine all schools of rent theory within the debate among neoclassical economists. See Fine (1983) for a criticism of Buchanan on this score and for a more detailed consideration of the debate over rent theory.
11. Wessel implicitly agrees with this assessment of the link between a price-determined rent and a partial equilibrium analysis; he also recognizes that the concept of general equilibrium was not an immediate and universal result of marginalism:

> In reality, of course, economists have long known that rent is neither price determining nor price determined since neither rent nor price is a basic determinant of the system. . . . We know from Cassel's (1932) simple general equilibrium model that these forces are the conditions of supply of the agents, the technical coefficients, and the preference patterns of consumers.

12. Steele (1967) argues that Marshall is hinting at a mineral replacement cost to restore the destructible powers of the soil (e.g. by the cost of discovery of new reserves or alternative energy sources) in his reference to the excess of receipts over production costs only being due to royalties 'in part'. This is clearly wrong (the other part of the surplus referred to by Marshall is the difference in costs of extraction relative to the margin) and reflects Steele's imposition of an interpretation based on the modern preoccupation with the renewal of exhaustible resources on to Marshall's analysis.
13. It is significant that for general equilibrium theory producer surplus has to be abandoned as a concept. See Mishan (1968).
14. Gray also introduces the innovation of discounting the mineral according to date of extraction, a procedure more associated with Hotelling (1931) and now taken up by modern theories of exhaustible resources.
15. See also the review by Devarajan and Fisher (1981) of the literature on exhaustible resources following Hotelling fifty years on. A royalty is defined as a price net of extraction and is without discussion of the role of landed property.

Notes

Chapter 4 Royalties: from private obstacle to public burden?

1. This chapter is primarily based on Fine (1984) and Brunskill *et al.* (1985).
2. This was the New Domesday Survey. See Bateman (1883) for a presentation and correction of much of its results. If Scotland is anything to judge by only a little has changed in the passage of one hundred years, and this is mainly through the buying up of large estates by multinational, institutional investors. See McEwen (1977). Royalty ownership was even more concentrated, as revealed in the next section,.
3. See Arnot (1949) for the narrative contained in the rest of this paragraph.
4. The MFGB was the last major union to support Labour MPs exclusively, for example. See Gregory (1968).
5. Most of the miners' evidence is to be found in the Second Volume of Minutes of Evidence.
6. See Appendix IV to the Final Report.
7. See the evidence of Wilson, Small, Cowey, Aspinwall, and Woods. Keir Hardie believed that royalties only had an indirect effect on wages through downward labour market pressure caused by higher unemployment brought about by lost export markets.
8. See the evidence of Small.
9. See the questioning of Aspinwall.
10. Those against state ownership included Aspinwall, Brown, Cowey, and Haslam.
11. An exception is to be found in Smillies' argument that some of the costs saved by more efficient pumping could find their way into the pockets of the miners.
12. Calculated from a table in Minutes of Evidence, p. 230.
13. The greater number of claimants as compared to those found in income tax returns is perhaps unsurprising given greater desire to claim compensation than to pay taxes (although compensation is for potential and hence future revenue).
14. Details such as those given here for Scotland are also calculable for other areas from data collected by the author for compensation paid upon nationalization of the royalties. The data cover information by individual mine, royalty owner and district, these in turn divided down by county.
15. This trend is confirmed by Church (1986: 12) for the UK more generally with the estimate that landowners were responsible themselves for as much as 10–15 per cent of output in 1830, half that by the 1870s, and a negligible proportion by 1913.
16. The figure for 1938 is calculated from the data for compensation paid for nationalization of the royalties.
17. The CMRC and Gowers in particular got off to a very bad start. Transferred to the CMRC in 1930 from the Inland Revenue, after previously having been Head of the Mines Department at the Board

of Trade from 1921 to 1927, his salary leapt from £3,000 to £7,000 per annum and was the object of Parliamentary dispute (Thomas 1937), no doubt inspired by parliamentary representation of the mine owners' opposition to the CMRC's cause. See also Supple (1986). In retrospect, Gowers was to offer a bitter indictment of the mine owners and the obstruction to the CMRC's work in his evidence to the Coal Industry Nationalization Tribunal in 1946, Public Record Office, POWE, 28/108. In the war he served as High Commissioner for London, justifying the appearance of his portrait in the Imperial War Museum. But he is best known for his guide for civil servants in the art of writing plain English (Gowers 1948), which still remains a standard text.
18. Much of this account is taken from Rooke (1937).
19. See Fine (1979) for an exposition of Marx's theory of agricultural rent, and also the debate with Ball (1980), all reproduced in Fine (1986).
20. For a critical analysis of the structure of housing provision see Ball (1983) and for a critique of its political reduction to the dimensions of tenure/distribution, Ball (1985).

Chapter 5 Returning to factor returns: the late nineteenth century coal industry

1. This chapter draws heavily upon Fine (1982c).
2. See Fine (1982b and 1983) and chapters 3 and 4 for the issue of rent in the context of coal mining.
3. See, for example, Lazonick (1981). In later work, Lazonick (1983) also emphasizes vertical integration into weaving and other management factors.
4. The arguments here complement many of those of Nicholas (1982).
5. See Church (1986: 116) for whom 'the overwhelming importance of fixed capital, however defined, is indisputable'.
6. In this context an even more inappropriate application of production function estimation is made by Jones (1983) for the South African coal industry – where, just for starters, competitive labour markets with wages equating marginal products are notably absent! For a critique, see Fine (1989b).
7. This has been laid out in the simplest terms in Fine (1980).
8. See, for example, Bliss (1975).
9. These comments, of course, apply to all production function measurement which uses factor rewards, and not simply to McCloskey. The best economists and econometricians recognize this and attempt to estimate general equilibrium systems. Whether this is a worthwhile task is another matter altogether.
10. See chapter 4 for greater details of distribution of royalty ownership.
11. See chapter 4 for some discussion of decline of landowners as mine owners.

Notes

Chapter 6 Returns to scale in the interwar coal industry

1. This chapter is based on Evans and Fine (1980a).
2. Buxton agrees that reorganization of underground haulage was desirable. That this would itself be subject to economies of scale seems most likely.
3. Note that no attempt is made to enter into the debate over the meaning of amalgamation, rationalization, etc. between Buxton (1972a and b), Kirby (1972), and Johnson (1972). It is not clear that these distinctions can be usefully made in coal-mining where, for example, so much uncertainty is associated with production so that multiple ownership of mines allows faces or pits to be held in reserve according to the vagaries of either market demand or geological faulting. Note also that mechanization will usually refer to coal cutting by machine.
4. The eighteen districts are listed in appendix 6.1. They are formed from the twenty-five districts used in HM Inspectors Reports. This was necessary as separate data are unavailable for all years, since aggregation across some districts is used in the official statistics for certain years.
5. In particular, the use of an average representative mine may conceal important relations between the size distribution of mines and productivity.
6. Rhodes (1945) uses a similar method to the one proposed here but with less sophisticated techniques. But he sets $\gamma = 1$ and imposes constant returns to scale, $\alpha + \beta = 1$. Ingeniously, he attempts to test for economies of scale, having already ruled them out, by comparing summer with winter output on the presumption that fixed capital is more fully used in the short run in winter. He finds evidence of increasing returns.
7. See the discussion and references in the previous chapter on the problems of measuring total factor productivity.
8. This is a method suggested by Maddala (1977). It presumes that there are uniform characteristics, in this case a and c, across districts, but the estimate of these is distorted by other district-specific properties. The more these intervene, as measured by variance, the less the district estimate should count.
9. This is not as bad as it sounds, even if there is an assumption of uniform productivity increase, since the index was constructed before taking account of mine size and labour inputs.

Chapter 7 The diffusion of mechanical coal cutting

1. This chapter is predominantly based upon Evans and Fine (1980b) and Fine (1989f).
2. The eighteen districts are formed from the twenty-five distinguished in the reports of HM Inspector of Mines. The reduction to eighteen districts is necessary because of the aggregation across certain

districts in these reports for certain years. The districts are identified in table 7.1 after which they are numbered from one to eighteen.
3. See Greasley (1979: 34) who argues that 'with the introduction of Mayor and Coulson's "Samson" chain coal cutter in 1930, and Anderson Boyes A.B. 15 chain machine in 1930 the design of Longwall undercutting machines was approaching a steady state'. See also Appendix 6.3.
4. It is possible for the actual and estimated initial levels to diverge.
5. The two methods are derived according to whether movements in p or $1/p$ are estimated where p is the proportion of mechanization.
6. The fact that the percentage of mechanized mines was 44 per cent by 1938 suggests that according to this fitting of the diffusion process, mechanization was more or less completed by 1938.
7. For a discussion of these problems, see Sen (1973).
8. For the classic estimation of a diffusion process, see Griliches (1957).

Chapter 8 The commanding heights of public corporation economics

1. The chapter draws upon parts of Fine and Harris (1987) and on Fine (1989d and 1989e).
2. See Privatisation (1986) for a global survey. For the Adam Smith Institute (1986), 'Dozens of countries have sent representatives to Britain to see the results at first hand, and to learn how it is done'. See also Aylen (1988).
3. Details of sales are to be found in Price-Waterhouse (1987). See also, for example, Bishop and Kay (1988) and Buckland (1987).
4. 500 staff meetings, 30 lawyers, 25 civil servants, 6 management briefings, 3 Queen's Counsel, 2 Secretaries of State, 1 Act of Parliament.

 This just for a management buy-out; quoted from the Chairman of NFC Ltd and cited in McLachlan (1983: 76).
5. Consider the heart-searching over defence and the appointment of Peter Levene to procurement in particular, Defence Select Committee (1984/85 and 1987/88).
6. Nevertheless, empirical work is retained because of the unpredictability of behaviour and to 'suggest hitherto unnoticed flaws or omissions or undeveloped propositions in economic theory', p. 26.
7. For an orthodox survey of the theory of property rights, see Furnbotn and Pejovich (1972). For treatment within an institutional tradition, see Williamson (1985). Buchanan (1986) provides an account within a broad framework from the right-wing Virginian school.
8. Littlechild (1981) has also defended monopolies against the charge of inefficiency through the pursuit of monopoly profits.

Notes

9. See also Kay's (1987) introduction to a group of papers on privatization, with his observation that five years earlier their content would have been entirely different.
10. The rapid rise of the new public sector economics is symbolized by the early appearance of survey articles. A simple account of the propositions concerned is to be found in Hemming and Mansoor (1988). An early contribution is provided by Kay and Silbertson (1984). See, however, Domberger and Piggott (1986) who argue that Kay and Thompson (1986) provides 'the most authoritative international contribution'.
11. For consideration of new developments in industrial economics, see Schmalansee (1988) and, often in a policy context, Mayer (1985) and Vickers (1985). It is significant that the more longstanding approaches to industrial economics, such as market and hierarchies and Coasian theory, have either been unable to respond to privatisation or have been elbowed aside by the synthesis.
12. This is, of course, no novelty as the presumption of perfect regulation to correct imperfect markets has long been criticized. See Schmalansee (1979) and Rees (1984), for example.
13; See Helm, Kay, and Thompson (1988: 47) who also point to the changing incidence of monopoly over time.
14. For a critique of contestability, see Shepherd (1984), Weitzman (1983), and Schwartz and Reynolds (1983) and Baumol (1983) for a response to the latter two. See also Baumol *et al.* (1986) and Brock (1983).
15. A partial equilibrium model necessarily excludes the issue of dynamics and stability and this reflects the limitations of the synthesis' approach . See Sharkey (1982) for a consideration of stability in the presence of natural monopoly.
16. The same is also true of the theory of contestability. In principle any economic agent is able to make a 'dawn raid' on a monopoly enjoying excess profits.
17. For an account of adverse changes in industrial relations following privatization, see TUC (1986).
18. See Bishop and Kay (1988) for an account of the rapidly rising salaries of management in privatized corporations. See also TUC (1987).
19. Evidence for the significance of ownership in changing the motivation and not just the constraints of economic agents is to be found in the ESOP and MBO literature, on which see Green (1988) and Estrin *et al.* (1988), for example. For a critique of the associated notion of people's capitalism, see Fine (1988b).
20. This would appear to follow from Helm's (1986) view of the relationship between the state and the economy. For him, following on essentially from a distinction between positive and normative economics, the state is justified in intervening when this will improve welfare (and there can be no general presumption in favour of market provision). This even-handed outlook, however,

Notes

presumes that the state is in principle separate from the economy in the first instance and a choice has to be made of how much this should remain so. Even in the most primitive (exchange) economies, the state is of necessity intervening – through property laws, ratifying the form of money, etc. For an implicit critique of the Helm position from a Marxist stance, seeing it as reflecting a 'real illusion' of the separation of the economy from society, see Holloway and Picciotto (1978).

21. See Helm, Kay, and Thompson (1988: 44) and Molyneux and Thompson (1987: 73). Direct provision is warranted if a minimum standard of consumption is absolutely required.
22. For conflict and power as alternative themes in studying nationalized industries, see Taylor (1984).
23. See Millward (1982), Pryke (1982), Helm, Kay, and Thompson (1988), Molyneux and Thompson (1987), Hammond *et al.* (1986), and Kay and Thompson (1986) for some assessment of the performance of nationalized industries. Over recent years, the productivity 'miracle' of the British economy has been associated with a greater than average performance of the public corporations. Privatization as such does not seem to have been significant in improving performance, see Bishop and Kay (1988). For an assessment of the par performance of the privatized British Telecom as compared to its European state-owned counterparts, see Foreman-Peck and Manning (1988).
24. See Kay, Mayer, and Thompson (1987), Thompson (1987: 368), Jaffer and Thompson (1987), Hammond *et al.* (1985) and Kay and Thompson (1987: 82) for discussion of incumbents. Note that this implies the rejection of (the relevance of) contestability theory.
25. See Yarrow (1986: 8 and 52) for a recognition of the need for a case-by-case approach within a set of broad analytical themes.
26. See Helm, Kay and Thompson (1988: 49), Molyneux and Thompson (1987: 72) and Yarrow (1986: 324) for the erroneous view that historically nationalizations arose out of the incidence of natural monopoly.
27. This has been most forcefully argued by Lazonick (1981), for example, as against the total factor productivity approach. See the surrounding debate with Sandberg.
28. The predominant macroeconomic concern has been with the budgetary effects of the short-term cash gains from selling assets.
29. Interestingly, Cairns and Mahabir (1988), in rejecting the Baumol *et al.* approach for a single sector, put forward the idea that contestability would be more appropriate where sunk capital can be easily switched through diversification into related sectors.
30. For a recent attempt to construct industrial policy in a much broader framework of economic and social factors, see the GLC's (1984b) industrial strategy.
31. See especially Helm and Yarrow (1988) and Kay and Vickers (1988).

Notes

32. See Schmalansee (1979), Joskow and Schmalansee (1986), and Stelzer (1988) who describes regulation as a new British import.
33. Again, just as for oligopoly theory there is a duality between Cournot and Bertrand type analysis.
34. Note that one of the perverse consequences of the regulation literature is to provide guidance for the private sector on how to undermine regulatory intent.
35. Consider, for example, Brown and Sibley (1986) which is almost exclusively concerned with efficiency pricing in the context of the static maximization of consumer surplus.

Chapter 9 Privatization and property rights: from electricity to coal

1. This chapter draws in part from Fine (1989c). See also Fine (1984).
2. See the evidence presented to the Energy Select Committee (1987/88a) by the ESTUC.
3. See Glyn (1988) and also GLC and NUM evidence to the Sizewell Inquiry (Layfield 1986).
4. There are also the health costs of unemployment. For a review of the issues and evidence, see Smith (1987).
5. In the cases of the colliery reviews at Bates and Cadeby, the independent inspectors asked British Coal to reconsider their closure decisions in view of the wider economic and social costs.
6. See Henderson (1977) for an early contribution. In the light of the protection to be afforded the nuclear industry after privatization, its excessive costs have at last received the attention they deserve from orthodox economists. See Helm (1987) and Yarrow (1988a), for example. Bunn and Vlahos (1989) calculate the extra cost of the nuclear constraint in providing for energy diversity.
7. See, for example, Sedgemore (1980) and Sweet (1985) for an account of the power of the nuclear establishment and Fine and Harris (1985) for an attempted explanation for it.
8. The issues of cost and safety are not, however, independent of each other. See Fine and Harris (1985) for the suggestion that there is a 'nuclear cycle' in which the mounting safety implications derived from the operating experience of one generation of stations leads to corresponding increases in capital costs in the next, as modifications are made to accommodate what has been learnt from operating experience.
9. For the classic critique of UK R&D policy, see Maddocks (1983).
10. See, for example, Vickers and Yarrow (1985) and EPEA (1987).
11. See, for example, Henney (1987), Holmes *et al.* (1987), and Yarrow (1988b).
12. Throughout discussion will be primarily concerned with privatization of electricity in England and Wales and not with Scotland, for which see Industry Department for Scotland (1988), nor Northern

Ireland. For a critique of the plans for Scotland, see Kerevan and Saville (1987).
13. See especially many of the Memoranda submitted to the Energy Select Committee (1987/88a) and Helm (1988).
14. There was some discussion over the meaning of the term 'spatchcock', but it was generally taken to be derogatory. The merit order seeks to ensure that power stations are used to meet electricity demand in the sequence of their economic efficiency.
15. See chapter 8 and also Redwood (1988: 31):

> First of all, when you are setting about your task of privatisation, don't go for the two-year slog or the two million pound consultancy job to work out absolutely every detail . . . if you are seriously interested in privatising, there is no substitute for doing it.

Interestingly, the Energy Select Committee (1987/88b) report that £6 million had already been spent on advising the Government on privatization of electricity.
16. See also the Open University Energy and Environment Group's Memorandum 87 to the Energy Select Committee (1987/88a) and Chesshire (1986).
17. See also GLC (1983), Feickert (1985), and Rudig (1986).
18. As described in Energy Select Committee (1980/81). See also Fine and Harris (1985).
19. On power station construction, a history of its problems, and the sequence of reports upon them, see Monopolies and Mergers Commission (1980/81). Since then, construction has been limited by excess capacity, but see also Chesshire (1987) for a discussion in the context of privatization.
20. See Harlow (1987) for a more general discussion of the role of the public sector on the private sector in this context.
21. As laid out in a series of reports by NEDO, the most recent being NEDC (1988) and McKinsey (1987). See also Fine and Harris (1985), and Walker (1986) specifically for the relation between IT and electricity supply. Many of the memoranda to the Energy Select Committee (1987/88a) from industry express concern about continuing R&D support and co-ordination.
22. For some discussion of ballotitis, see Hyman (1986) and for a less critical assessment of not holding a ballot, Goodman (1985). On policing, see later.
23. For the impact of imports, following privatization of electricity, see Gladstone and Dewhurst (1988), and for the impact on the coal machinery sector, BLMA (n.d.).
24. Thus, Jones (1988), the UKAEA's Chief Economic Adviser, argues for continuing nuclear research funding on the grounds of 43,000 jobs created directly (10,000 in power stations) and 177,000 indirectly. He also suggests that in the absence of the nuclear

Notes

   power programme, coal prices would have been higher because of
   the need to employ more marginal mines. This is fallacious as it
   presumes increasing long-run costs of producing coal.
25. See McCloskey's Memorandum 82 to the Energy Select Committee
    (1986/87b), for example.
26. See also Newberry (1986: 32):

   > The price actually paid for the coal is in a sense a relatively
   > minor issue, as it is a straight transfer payment from one
   > nationalised industry to another – the question will be who
   > should show the losses arising, and how should they be
   > financed. Provided the actual price paid is seen in this way, and
   > has no consequential effects on future production decisions, then
   > little is at stake.

   This is not, however, what has happened.
27. For an ironic use of the necessity of writing off sunk costs in
    making decisions about the future see Minford and Kung (1984)
    who rallied behind the Government by pointing out that the costs
    incurred by the strike by week 34 (including those of policing)
    could not be retrieved by throwing in the towel. They also argue,
    as leading monetarists, that the calculations of the costs of
    unemployment of Davies and Metcalf (1984), let alone the more
    extreme views of Glyn (1984), are inappropriate. This is because
    closures of uneconomic pits will increase employment by releasing
    greater resources into the rest of the economy.
28. Note also that closures have often entailed the abandonment of
    mines that have had recent major investments. This suggests a
    sudden strategy of capacity limitation or that British Coal's invest-
    ment programme has been irrational. A kinder interpretation is that
    investment funds have been wasted in the futile attempt to keep
    uneconomic pits open (rather than their being starved of funds to
    ensure that they close).
29. What follows draws heavily on just a few of the arguments of
    Berry *et al.* (1985a). For a more extended description of the British
    Coal Accounts in operation, see Berry *et al.* (1985b). The
    accounting procedures of British Coal were also subject to
    criticisms by the Monopolies and Mergers Commission (1983) and
    these are heavily endorsed by Kerevan and Saville (1985a) and
    Cutler *et al.* (1985). British Coal took the unprecedented step of
    appointing a committee, Custis *et al.* (1985), to inquire into the
    allegations made in Berry *et al.* (1985a). They found that whether a
    pit was closed or not depended upon a number of flexible criteria.
    See also Cooper and Hopper (1988).
30. On welfare legislation and the miners, see Jones and Novak (1985),
    Evans *et al.* (1985), Rose (1985), and Booth and Smith (1985).
31. See Sweet (1985) and Cutler *et al.* (1985). See also NCB (1984).
32. In the vernacular of game theory, the separate area unions were

caught in a prisoners' dilemma or, more exactly, an assurance game if it is assumed that all benefit most if all are out on strike.
33. See Scraton (1985: 256/7) who disagrees with Kettle's (1985a) assessment of the continuing local police autonomy respected by the NRC.
34. Figures from an answer to a question in the House of Lords on 18 March 1985, Hansard, 389–90. See also Percy-Smith and Hillyard (1985).
35. See also the NCCL (1984) critique of the police, even if with support of the rights of scabs to work (at policing costs all too clear to see – 2000 police for one scab at Cortonwood).
36. See Lloyd (1985) and GLC Police Committee Reports, PC 231, July 1984 and PC 267, October 1984.
37. See Geary (1985) for an account of policing of industrial disputes over the last century.
38. For an account of the construction of the ideology of mugging and for its deconstruction, see Hall *et al.* (1978). A more general account of the ideology of criminalization is given by Pearson (1983). See also Gordon (1985) and Bunyan (1985).
39. For some discussion of the ideology covered in this paragraph, see Williams (1985) and Samuel (1985).
40. For some account of police violence, see Coulter *et al.* (1984). Here, a full account cannot be given of the policing of the strike and its implications. Other aspects include the use of the National Computer Centre to register pickets' cars, the harassment of private bus companies hired by miners, police tactics in the use of handcuffs, truncheons, horses, dogs, photographs, *agents provocateurs*, telephone tapping, plea bargaining, and the denial of access to solicitors. See Welsh Campaign for Civil Liberties (1985).
41. By the end of the strike, 44 per cent of those questioned were fairly or very sympathetic to the miners, three-quarters of Labour voters and even one-quarter of Tory voters, Cumberbatch *et al.* (1986).
42. Thus, Cumberbatch *et al* (1986) report that Scargill was four times more likely to be interviewed critically on the television than representatives of British Coal, and presentation of the issues was with the official view predominant, with some exceptions in documentaries as opposed to news programmes. See also Jones *et al.* (n.d.).
43. See Howell (1985), for example.

Chapter 10 Coal: the ultimate privatization

1. For an earlier presentation, see Fine and O'Donnell (1981).
2. These propositions are particularly apposite for the coal industry, see Fine *et al.* (1985a and 1985b).
3. He is better known for being one of the founding fathers, there being few mothers, of the marginalist revolution. His son, though

Notes

less well known, contributed a much better and thorough understanding of the industry, Jevons (1915).
4. It is also worth observing the change of management attitudes in the industrial relations arena, with hardliners to be favoured at all levels and demoralization and departure for the rest. This view depends predominantly upon anecdotal evidence, but see Gibbon (1988) and also MacGregor (1986) as chief instigator.
5. See especially Burns et al. (1985) and Winterton (1985), but also Fine and Harris (1985) on new technology. For latest developments, see Winterton (1988), and for their relation to the 1984/85 strike, Winterton and Winterton (1989).
6. For a consideration of the private mining companies in the context of privatization, see Kleinwort-Grieveson (1988).
7. For studies of Anglo-American, see Innes (1983) and Pallister *et al.* (1987).
8. See also Robinson (1988). Dieter Helm, as reported in the *Economist* of 12 November 1988, suggests a break-up into area companies with opencast being used as a 'sweetener' to make the less profitable areas more attractive. It is not clear, however, why uneconomic pits would be retained after privatization.

References

Adam Smith Institute (1986) *Privatisation Worldwide*, London: ASI Research.
Adeney, M. and J. Lloyd (1986) *Loss without Limit: The Miners' Strike of 1984/5*, London: Routledge and Kegan Paul.
Aldcroft, D. (ed.) (1968) *The Development of British Industry and Foreign Competition, 1875-1914: Studies in Industrial Enterprise*, London: Allen and Unwin.
Arnot, R. (1949) *The Miners 1881-1910: A History of the Miners Federation of Great Britain*, London: Allen & Unwin.
Ascher, K. (1987) *The Politics of Privatisation: Contracting Out Public Services*, London: MacMillan.
Ashton, T. and J. Sykes (1929) *The Coal Industry of the Eighteenth Century*, Manchester: Manchester University Press.
Ashworth, W. (1986) *The History of the British Coal Industry, 1946-82: the Nationalised Industry*, V. Oxford: Clarendon.
Aston, T. and C. Philpin (eds) (1985) *The Brenner Debate*, Cambridge: Cambridge University Press.
Aylen, J. (1988) 'Privatising in Developing Countries', in Johnson (ed.) (1988).
Baker, J. (1989) 'Business Objectives and Organisation', Notes for National Power Managers Meeting, 25 January, London, mimeo.
Ball, M. (1980) 'On Marx's Theory of Agricultural Rent: a Reply to Ben Fine', *Economy and Society*, 9, 3: 304-26, reproduced in Fine (ed.) (1986).
Ball, M. (1983) *Housing Policy and Economic Power: the Political Economy of Owner Occupation*, London: Methuen.
Ball, M. (1985) 'Coming to Terms with Owner Occupation', *Capital and Class*, 24, Winter: 15-44.
Ball, M. (1988) *Rebuilding Construction: Economic Change in the British Construction Industry*, London: Routledge.
Ball, M. *et al.* (eds) (1984) *Land Rent, Housing and Urban Planning: a European Perspective*, London: Croom Helm.
Baran, P. and P. Sweezy (1968) *Monopoly Capital*, Harmondsworth: Penguin.
Bateman, J. (1971) *The Great Landowners of Great Britain and Ireland*,

References

fourth edition, London: Harrision, 1883, reprinted by Leicester University Press, with a commentary by D. Spring.

Baumol, W. (1982) 'Contestable Markets: An Uprising in the Theory of Industry Structure', *American Economic Review*, 72, 1: 1-15.

Baumol, W. (1983) 'Contestable Markets: An Uprising in the Theory of Industry Structure: Reply', *American Economic Review*, 73: 491-6.

Baumol, W. et al. (1982) *Contestable Markets and the Theory of Industry Structure*, New York: Harcourt Brace Jovanovich.

Baumol, W. et al. (1986) 'Contestability: Developments since the Book', Oxford Economic Papers, 38: 9-36.

Beesley, M. and S. Littlechild (1983) 'Privatisation: Principles, Problems and Priorities', *Lloyds Bank Review*, July, reprinted in Kay et al. (1987).

Benwell CDP (1979) *The Making of a Ruling Class*, Newcastle upon Tyne: Benwell Community Project.

Berry, T. et al. (1985a) 'NCB Accounts – a Mine of Mis-Information', *Accountancy*, 96, 1097, January: 10-2.

Berry, T. et al. (1985b) 'Management Control in an Area of the NCB: Rationales of Accounting Practices in a Public Enterprise', *Accounting, Organizations and Society*, 10, 1: 3-28.

Beynon, H. (ed.) (1985) *Digging Deep*, London: Verso.

Beynon, H. et al. (1986) 'Nationalised Industry Policies and the Destruction of Communities: Some Evidence from North East England', *Capital and Class*, 29, Spring: 27-57.

Bishop, M. and J. Kay (1988) *Does Privatisation Work?: Lessons from the UK*, London: Centre for Business Strategy, London Business School.

Blake, N. (1985) 'Picketing, Justice and the Law, in Fine and Millar (eds) (1985).

Blanden, M. (1988) 'Murder on the Privatisation Express', *Banker*, 138, April: 22-4.

Bliss, C. (1975) *Capital Theory and the Distribution of Income*, Amsterdam: North Holland.

BLMA (n.d.) *The Case for Britain's Coal*, London: British Longwall Mining Association.

Board of Trade (1931-40) *Reports Under Section 7 of the Coal Mines Act 1930. On the Working of the Coal Selling Schemes Under Part I of the Act*, Annual and Occasionally Quarterly Reports with various Command numbers, London: HMSO.

Booth, A. and R. Smith (1985) 'The Irony of the Iron Fist: Social Security and the Coal Dispute of 1984-85', in Scraton (ed.) (1985),

Boyfield, K. (1985) *Putting Pits back into Profit*, London: Centre for Policy Studies.

Boyns, T. (1987) 'Rationalisation in the Interwar Period: the Case of the South Wales Steam Coal Industry', *Business History*, XXIX, 3: 282-303.

Braverman, H. (1974) *Labour and Monopoly Capital: the Degradation of Work in the Twentieth Century*, New York: Monthly Review Press.

Brenner, R. (1977) 'The Origins of Capitalist Development: a Critique of

References

Neo-Smithian Marxism', *New Left Review*, 104, July/August: 25–93.

Breusch, T. and A. Pagan (1980) 'The Lagrange Multiplier Test and its Application to Model Specification in Econometrics', *Review of Economic Studies*, 47, 1, January: 239–53.

Brittan, S. (1986) 'Privatisation: a Comment on Kay and Thompson', *Economic Journal*, 96, March: 33–8.

Brock, W. (1983) 'Contestable Markets and the Theory of Industry Structure: a Review Article', *Journal of Political Economy*, 91: 1055–66.

Brown, A. (1987) 'Privatisation Pays Dividends for Accountants', *Accountant*, September: 6–7.

Brown, L. and S. Selzer (1985) *The Miners' Strike 1984–85: A Select Bibliography*, London: British Coal Library, mimeo.

Brown, L. and S. Selzer (1986) *The Miners' Strike 1984–85: A Select Bibliography, Supplement*, London: British Coal Library, mimeo.

Brown, S. and D. Sibley (1986) *The Theory of Public Utility Pricing*, Cambridge: Cambridge University Press.

Brunskill, I. et al. (1985) 'The Ownership of Coal Royalties in Scotland', *Scottish Social and Economic Review*: 78–89.

Buchanan, D. (1929) 'The Historical Approach to Rent and Price Theory', *Economica*, 9, June: 123–55.

Buchanan, J. (1986) *Liberty, Markets, State: Political Economy in the 1980s*, Brighton: Wheatsheaf.

Buckland, R. (1987) 'The Costs and Returns of the Privatisation of Nationalised Industries', *Public Administration*, 65, 3, Autumn: 241–58.

Bunyan, T. (1985) 'From Saltley to Orgreave via Brixton', in Scraton (ed.) (1985).

Bunn, D. and K. Vlahos (1989) 'Evaluation of the Nuclear Constraint in a Privatised Electricity Supply Industry', *Fiscal Studies* 10, 1, Feb: 41–52.

Burns, A. et al. (1985) 'The Restructuring of the British Coal Industry', *Cambridge Journal of Economics*, 9, 1: 93–110.

Butler, E. (ed.) (1988) *The Mechanics of Privatisation*, London: Adam Smith Institute.

Buxton, N. (1970) 'Entrepreneurial Efficiency in the British Coal Industry Between the Wars', *Economic History Review*, 23, 3: 476–97.

Buxton, N. (1972a) 'Entrepreneurial Efficiency in the British Coal Industry between the Wars: Reconfirmed', *Economic History Review*, 25, 4: 658–64.

Buxton, N. (1972b) 'Avoiding the Pitfalls. Entrepreneurial Efficiency in the Coal Industry Again', *Economic History Review*, 25, 4: 669–73.

Buxton, N. (1978) *The Economic Development of the British Coal Industry*, London: Batsford Academic.

Buxton, N. (1979) 'Coalmining', in Buxton and Aldcroft (eds) (1979).

Buxton, N. and D. Aldcroft (eds) (1979) *British Industry Between the Wars*, London: Scolar Press.

Byatt, I. (1985) 'Market and Non-Market Alternatives in the Supply of Public Services: British Experience with Privatisation', in Forte and Peacock (eds) (1985).

References

Cairns, R. and Mahabir, D. (1988) 'Contestability: a Revisionist View', *Economica*, 55, May: 269–76.

Callinicos, A. and M. Simons (1985) *The Great Strike*, London: SWP.

Carsberg, B. (1986) 'Regulating Private Monopolies and Promoting Competition', *Long Range Planning*, 19, 6, December: 75–82.

Cassel, G. (1932) *The Theory of Social Economy*, New York: Brace, Harcourt and Co.

Catalano, C. and D. Massey (1979) *Capital and Land*, London: Edward Arnold.

Cecchini, P. (1988) *The European Challenge, 1992: the Benefits of a Single Market*, Aldershot: Wildwood House,.

Chesshire, J. (1986) 'An Energy Efficient Future: a Strategy for the UK', *Energy Policy*, 14, 5, October: 395–412.

Chesshire, J. (1987) 'The Privatisation of the UK Electricity Supply – Introducing Competition?', Electricity Consumers' Council Conference on Privatisation, October, London: ECC.

Christian, L. (1985) 'Restriction Without Conviction', in Fine and Millar (eds) (1985).

Church, R. (1986) *The History of the British Coal Industry, Volume 3, 1830–1913: Victorian Pre-eminence*, Oxford: Clarendon.

Clarke, R. and T. McGuinness (eds) (1987) *The Economics of the Firm*, Oxford: Blackwell.

CMRC (1933) *Report to the Secretary for Mines*, Cmd 4468, London: Mines Department.

Coal Commission (1945) *Report to the Minister of Fuel and Power for the Twelve Months Ended 31st March 1945*, London: HMSO.

Cooper, D. and T. Hopper (eds) *Debating Coal Closures: Economic Calculation in the Coal Dispute 1984–5*, Cambridge: Cambridge University Press.

Coulter, J. *et al.* (1984) *State of Siege*, London: Canary.

Crafts, N. (1987) 'Cliometrics, 1971–1986: a Survey', *Journal of Applied Econometrics*, 2, 3, July: 171–92.

Crafts, N. (1988) 'Privatisation', *Management Today*, August: 85.

Crick, M. (1985) *Scargill and the Miners*, Penguin: Harmondsworth.

Cromar, P. (1977) 'The Coal Industry on Tyneside 1771–1800: Oligopoly and Spatial Change', *Economic Geography*, 53: 79–94.

Cumberbatch, G. *et al.* (1986) *Television and the Miners' Strike*, London: Broadcasting Research Unit.

Currie, D. and R. Smith (eds) (1981) *Socialist Economic Review*, 1, London: Merlin Press.

Custis, P. *et al.* (1985) *Report of an Independent Committee of Enquiry on Certain Accounting Matters Relating to the Affairs of the National Coal Board*, London: NCB.

Cutler, T. *et al.* (1985) *The Aberystwyth Report on Coal*, Aberystwyth: University College of Wales.

Dasgupta, P. and G. Heal (1979) *The Economics of Exhaustible Resources*, Cambridge: CUP.

Davies, G. and D. Metcalf (1984) *Pit Closures: Some Economics*, London: London Weekend Television.

References

Defence Select Committee (1984/85) *The Appointment and Objectives of the Chief of Defence Procurement*, HC 430, London: HMSO.

Defence Select Committee (1987/88) *Business Appointments: the Acceptance of Appointments in Commerce and Industry by Members of the Armed Forces and Officers of the Ministry of Defence*, HC 392, London: HMSO.

Department of Energy (1988) *Privatising Electricity. The Government's Proposals for the Privatisation of the Electricity Supply Industry in England and Wales*, Cmd 327, London: HMSO.

Devarajan, S. and A. Fisher (1981) 'Hotelling's "Economics of Exhaustible Resources": Fifty Years Later', *Journal of Economic Literature*, 19, March: 65–73.

Dick, B. (1987) *Privatisation in the UK: the Free Market versus State Control*, York: Longmans.

Dilnot, A. and D. Helm (1987) 'Energy Policy, Merit Goods and Social Security', *Fiscal Studies*, 8, 3: 29–48.

Dintenfass, M. (1984) 'The Politics of Producers' Cooperation: The FBI-TUC-NCEO Talks, 1929–33', in Turner (ed.) (1984).

Dintenfass, M. (1985) 'Industrial Decline: Four British Colliery Companies between the Wars', unpublished PhD Thesis, Colombia University.

Dintenfass, M. (1988) 'Entrepreneurial Failure Reconsidered: the Case of the Interwar British Coal Industry', *Business History Review*, 62, Spring: 1–34.

Dobb, M. (1943) Review of Sweezy (1942), *Science and Society*, 7, 3.

Domberger, S. and J. Piggott (1986) 'Privatization and Public Enterprise: a Survey', *Economic Record*, 62, 177: 145–62.

Dornbusch, R. and R. Layard (eds) (1987) *The Performance of the British Economy*, Oxford: Clarendon Press.

Economist (1988) 'A Blueprint for Coal', 12 November.

Eldon Barry, E. (1965) *Nationalisation in British Politics*, London: Jonathan Cape.

Energy Economist (1988a) 'Nuclear Costs are Rising Exponentially', May: 2–5.

Energy Economist (1988b) 'Nuclear Power World Status Report', January: 15–20.

Energy Select Committee (1980/81) *The Goverment's Statement on the New Nuclear Power Programme*, HC 114, London: HMSO.

Energy Select Committee (1981/82) *Energy Conservation in Buildings*, HC 401, London: HMSO.

Energy Select Committee (1985/86) *The Coal Industry*, HC 196, London: HMSO.

Energy Select Committee (1986/87a) *The Coal Industry*, HC 176, London: HMSO.

Energy Select Committee (1986/87b) *The Government's Response to the Committee's First Report of Session 186/7, on the Coal Industry, HC 176*, HC 387, London: HMSO.

Energy Select Committee (1987/88a) *The Structure, Regulation and Economic Consequences of Electricity Supply in the Private Sector*, Third Report, HC 307, HMSO: London.

References

Energy Select Committee (1987/88b) *The Department of Energy's Spending Plans, 1988-89*, Fourth Report, HC 513, HMSO: London.

EPEA (1987) *Why the CEGB Should not be Broken up*, London: Engineers' and Managers Association.

Estrin, S. et al. (eds) (1988) 'Employee Share Ownership, Profit Sharing and Participation', Special Issue of *International Journal of Industrial Organization*, 6, 1, March: 1-154.

Evans, J. et al. (1985) 'Women and the Strike: It's a Whole Way of Life', in Fine and Millar (eds) (1985).

Evans, T. and B. Fine (1980a) 'Economies of Scale and the British Interwar Coal Industry', *Birkbeck Discussion Paper*, 76.

Evans, T. and B. Fine (1980b) 'The Diffusion of Mechanical Coal Cutting in the British Interwar Coal Industry', *Birkbeck Discussion Paper*, 75.

Fare, R. et al. (1985) 'The Relative Performance of Publicly-Owned and Privately-Owned Electric Utilities', *Journal of Public Economies*, 26: 89-106.

Feickert, D. (1985) 'Towards a New Future: Campaigning for Coal', in Beynon (ed.) (1985).

Fine, B. (1978) 'Royalties and the Interwar Coal Industry', *Birkbeck Discussion Paper*, 62.

Fine, B. (1979) 'On Marx's Theory of Agricultural Rent', *Economy and Society*, 8, 3: 241-78, reproduced in Fine (ed.) (1986).

Fine, B. (1980) *Economic Theory and Ideology*, London: Edward Arnold.

Fine, B. (1982a) 'Landed Property and the Distinction Between Royalty and Rent', *Land Economies*, 58, 3: 338-50.

Fine, B. (1982b) *Theories of the Capitalist Economy*, London: Edward Arnold.

Fine, B. (1982c) 'Landed Property and the British Coal Industry prior to World War I', *Birkbeck College Discussion Paper*, 120.

Fine, B. (1983) 'The Historical Approach to Rent and Price Theory Reconsidered', *Australian Economic Papers*, 22, 40: 132-43.

Fine, B. (1984) 'Land, Capital and the British Coal Industry', in Ball et al. (eds) (1984).

Fine, B. (1985) 'The Future of British Coal', *Capital and Class*, 23, Summer: 67-82.

Fine, B. (ed.) (1986) *The Value Dimension: Marx versus Ricardo and Sraffa*, London: Routledge and Kegan Paul.

Fine, B. (1988a) 'The British Coal Industry's Contribution to the Political Economy of Paul Sweezy', *History of Political Economy* 20, 2, Summer: 235-50.

Fine, B. (1988b) 'Is There Such a Thing as "People's Capitalism"?', *World Marxist Review*, 31, 2: 129-36.

Fine, B. (1989a) 'Economies of Scale and a Featherbedding Cartel? A Reconsideration of the Interwar British Coal Industry', *Economic History Review*, forthcoming.

Fine, B. (1989b) 'Total Factor Productivity versus Realism: the Case of the South African Coal Industry', *Zimbabwe Journal of Economics*, forthcoming.

References

Fine, B. (1989c) 'Privatisation of the Electricity Supply Industry: Broadening the Debate', *Energy Policy*, forthcoming.

Fine, B. (1989d) 'Scaling the Commanding Heights of Public Sector Economics', *Cambridge Journal of Economics*, forthcoming.

Fine, B. (1989e) 'Denationalisation', in Green (ed) (1989).

Fine, B. (1989f) 'Mechanisation of Coal Cutting in the Interwar British Coal Industry', *The Journal of European Economic History*, forthcoming.

Fine, B. and L. Harris (1985) *The Peculiarities of the British Economy*, London: Lawrence and Wishart.

Fine, B. and L. Harris (1987) 'Ideology and Markets: Economic Theory and the "New Right" ', in Miliband *et al.* (eds) (1987).

Fine, B. and A. Murfin (1984) *Macroeconomics and Monopoly Capitalism*, Brighton: Wheatsheaf.

Fine, B. and K. O'Donnell (1981) 'The Nationalised Industries', in Currie and Smith (eds) (1981).

Fine, B. and K. O'Donnell (1985) 'The Nationalised Industries', in Fine and Harris (1985).

Fine, B. and R. Millar (eds) (1985) *Policing the Miners' Strike*, London: Lawrence and Wishart.

Fine, B. *et al.* (1985a) 'Coal After Nationalisation', in Fine and Harris (1985).

Fine, B. *et al.* (1985b) 'Coal Before Nationalisation', in Fine and Harris (1985).

Flinn, M. (1984) *The History of the British Coal Industry, vol 2, 1700–1830: The Industrial Revolution*, with the assistance of D. Stoker, Oxford: Clarendon Press.

Flux, A. (1923) *Economic Principles*, London: Methuen.

Foreman-Peck, J. and D. Manning (1988) 'How Well is British Telecom Performing? An International Comparison of Telecommunications Total Factor Productivity', *Fiscal Studies*, 9, 3, August; 54–67.

Forrest, R. and A. Murie (1988) *Selling the Welfare State: the Privatisation of Public Housing*, London: Routledge.

Forte, K. and A. Peacock (eds) (1985) *Political Economy and Government Growth*, Oxford: Blackwell.

Foster, J. (1986) *The Theory of Monopoly Capitalism*, New York: Monthly Review Press.

Furnbotn, E. and S. Pejovich (1972) 'Property Rights and Economic Theory: a Survey of Recent Literature', *Journal of Economic Literature*, 10: 1137–62.

Gaffney, M. (ed) (1967) *Extractive Resources and Taxation*, Madison: University of Wisconsin Press.

Geary, R. (1985) *Policing Industrial Disputes; 1893–1985*, Cambridge: CUP.

Gibbon, P. (1988) 'Analysing the British Miners' Strike of 1984/5', *Economy and Society*, 17, 2, May: 139–94.

Gladstone, B. and D. Dewhurst (1988) *Generating Jobs: Electricity, Linked*

References

Industries and Privatisation, Barnsley: Coalfield Communities Campaign.

GLC (1983) *The Energy Economy*, Economic Policy Group Strategy Document no. 5, London: Greater London Council.

GLC (1984a) *London in the Dark*, London: Greater London Council.

GLC (1984b) *The London Industrial Strategy*, London: Greater London Council.

Glyn, A. (1984) *The Economic Case against Pit Closures*, Sheffield: NUM.

Glyn, A. (1988) 'The Economic Costs of Imports', Evidence Submitted to the Committee Stage of the Humberside Ports Bill, mimeo.

Goodman, G. (1985) *The Miners' Strike*, London: Pluto.

Gordon, P. (1985) ' "If They Come in the Morning. . . . ": The Police, the Miners and Black People', in Fine and Millar (eds) (1985).

Gowers, E. (1948) *Plain Words: a Guide to the Use of English*, S. Greenbaum and J. Whitcut, revised edition, 1986, London: HMSO.

Gray, L. (1914) 'Rent under the Assumption of Exhaustibility', *Quarterly Journal of Economics* 28: 466–89.

Greasley, D. (1979) *The Diffusion of Technology: the Case of Machine Coal Cutting in Great Britain, 1900–1938*, Unpublished PhD thesis, Liverpool.

Greasley, D. (1982) 'The Diffusion of Machine Cutting in the British Coal Industry, 1902–1938', *Explorations in Economic History*, 19: 246–268.

Green, A. (1985) "Research Bibliography of Published Materials Relating to the Coal Dispute", in Scraton (ed.) (1985).

Green, F. (ed.) (1989) *The Restructuring of the UK Economy*, Brighton: Harvester, forthcoming.

Green, S. (1988) 'The Incentive Effects of Ownership and Controls in Management Buy-outs', *Long Range Planning*, 21, 1: 26–34.

Gregory, R. (1968) *The Miners and British Politics, 1906–14*, Oxford: Oxford University Press.

Grieve Smith, J. (ed.) (1984) *Strategic Planning in Nationalised Industries*, London: MacMillan.

Griliches, Z. (1957) 'Hybrid Corn: an Exploration in the Economics of Technical Change', *Econometrica*, 25, Oct: 501–22.

Hagen, K. (1979) 'Optimal Pricing in Public Firms in an Imperfect Market Economy', *Scandinavian Journal of Economics*, 81: 475–93.

Hall, S. *et al.* (1978) *Policing the Crisis – Mugging, the State, Law and Order*, London: MacMillan.

Hammond, E. *et al.* (1985) 'British Gas: Options for Privatisation', *Fiscal Studies*, 6, 4: 1–20.

Hammond, E. *et al.* (1986) 'Competition in Electricity Supply: Has the Energy Act Failed?', *Fiscal Studies*, 7, 1: 11–33.

Harlow, C. (1987) *Commercial Interdependence: Public Corporations and Private Industry*, London: Policy Studies Institute.

Hatcher, J. (n.d) *The History of the British Coal Industry*, vol 1, *Before 1700*, Oxford: Clarendon Press, forthcoming.

Hausmann, W. (1980) 'A Model of the London Coal Trade in the

Eighteenth Century', *Quarterly Journal of Economics*, XCIV, February, 1: 1-14.

Hausmann, W. (1984a) 'Cheap Coals or Limitation of the Vend? The London Coal Trade, 1770-1845', *Journal of Economic History*, XLIV, 2, June: 321-28.

Hausmann, W. (1984b) 'Market Power in the London Coal Trade: the Limitations of the Vend, 1770-1845', *Explorations in Economic History*, 21: 383-405.

Hay, D. (1987) 'Competition and Industrial Policies', *Oxford Review of Economic Policy*, 3, 3: 27-70.

Helm, D. (1986) 'The Economic Borders of the State', *Oxford Review of Economic Policy*, 2: i-xxiv.

Helm, D. (1987) 'Nuclear Power and the Privatisation of Electricity Generation', *Fiscal Studies*, 8, 4: 69-73.

Helm, D. (1988) 'Regulating the Electricity Supply Industry', *Fiscal Studies*, 9, 3: 86-105.

Helm, D., J. Kay, and D. Thompson (1988) 'Energy Policy and the Role of the State in the Market for Energy', *Fiscal Studies*, 9, 1: 41-61.

Helm, D. and G. Yarrow (1988) 'The Assessment: The Regulation of Utilities', *Oxford Review of Economic Policy*, 4, 2: i-xxxi.

Hemming, R. and A. Mansoor (1988) 'Privatisation and Public Enterprise', Occasional Paper, 56, Washington DC: IMF.

Henderson, D. (1977) 'Two British Errors: Their Probable Size and Some Possible Lessons', *Oxford Economic Papers*, 29, 2, July: 159-205.

Henley, A. (1988) 'Price Formation and Market Structure: the Case of the Inter-War Coal Industry', *Oxford Bulletin of Economics and Statistics*, 50: 263-78.

Henney, A. (1987) *Privatise Power. Restructuring the Electricity Supply Industry*, London: Centre for Policy Studies.

Hillman, M. (1984) *Conservation's Contribution to UK Self Sufficiency*, Energy Paper 13, London: Heinemann.

Hillman, M. and A. Bollard (1985) *Less Fuel, More Jobs: the Promotion of Energy Conservation in Buildings*, London: Policy Studies Institute.

Hilton, R. (ed.) (1978) *The Transition from Feudalism to Capitalism*, London: Verso.

Hobson, J. (1891) 'The Law of the Three Rents', *Quarterly Journal of Economics*, 5: 263-88.

Holloway, J. and S. Picciotto (eds) (1978) *State and Capital: a Marxist Debate*, London: Edward Arnold.

Holmes, A. (1988) 'Government White Paper with too many Blank Spaces', *Energy Economist*, 77, March: 5-9.

Holmes, A et al. (1987) *Power on the Market*, London: Financial Times Business Information.

Hope, C. (1987) 'What is the CEGB Worth? - The Impact of Future Government Action', *Management Studies Research Paper*, 4/87, Cambridge: University Engineering Department.

Hotelling, H. (1931) 'The Economics of Exhaustible Resources', *Journal of Political Economy*, 5: 263-88.

References

Howell, D. (1985) ' "Where's Ramsay McKinnock?": Labour Leadership and the Miners', in Beynon (ed.) (1985).

Hyman, R. (1986) 'Reflections on the Mining Strike', in Miliband *et al.* (eds) (1986).

Ince, M. (1988) 'Industrial Effects of UK Electricity Privatisation', *Energy Policy*, August: 409–14.

Industry Department for Scotland (1988) *Privatisation of the Scottish Electricity Industry*, Cmd. 327, Edinburgh: HMSO.

Innes, D. (1983) *Anglo American and the Rise of Modern South Africa*, London: Heinemann Educational.

Jaffer, S. and D. Thompson (1987) 'Deregulating Express Coaches: A Reassessment', *Fiscal Studies*, 7, 4: 45–68.

Jevons, H. (1915) *The British Coal Trade*, London: Kegan Paul Trench Trubner.

Jevons, W. (1865) *The Coal Question: an Inquiry Concerning the Progress of the Nation and the Probable Exhaustion of Our Coal Mines*, London: MacMillan.

Johnson, C. (ed.) (1988) *Privatisation and Ownership, Lloyds Bank Annual Review*, 1, London.

Johnson, W. (1972) 'Entrepreneurial Efficiency in the British Coal Industry Between the Wars: a Second Comment', *Economic History Review*, 25, 4: 665–68.

Jones, C. and T. Novak (1985) 'Welfare against the Workers: Benefits as a Political Weapon', in Beynon (ed.) (1985).

Jones, D. *et al.* (n.d.) *Media Hits the Pits: the Media and the Coal Dispute*, London: Campaign for Press and Broadcasting Freedom.

Jones, P. (1988) 'The Benefits of Nuclear Power', *Atom*, 379, May: 12–7.

Jones, R. (1983) 'Mechanisation, Learning Periods, and Productivity change in the South African Coal Mining Industry: 1950–80', *South African Journal of Economics*, 51, 4: 507–22.

Joskow, P. and R. Schmalansee (1986) 'Incentive Regulation for Electric Utilities', *Yale Journal on Regulation*, 4, 1: 1–49.

Kahn, E. (1987) 'The Effect of Conservation Programmes on Electricity Utility Earnings: Results of Two Case Studies', *Energy Policy*, 15, 3, June: 249–61.

Kay, J. (1987) 'Public Ownership, Public Regulation or Public Subsidy', *European Economic Review*, 31: 343–45.

Kay, J., C. Mayer, and D. Thompson (eds) (1987) *Privatisation and Regulation – the UK Experience*, Oxford: Clarendon Press.

Kay, J. and A. Silbertson (1984) 'The New Industrial Policy – Privatisation and Competition', *Midland Bank Quarterly Review*, Spring: 8–16.

Kay, J. and D. Thompson (1986) 'Privatisation: a Policy in Search of a Rationale', *Economic Journal*, 96, March: 18–32.

Kay, J. and D. Thompson (1987) 'Policy for Industry', in Dornbusch and Layard (eds) (1987).

Kay, J. and J. Vickers (1988) 'Regulatory Reform in Britain', *Economic Policy*, 7, October: 185–252.

Kerevan, G. and R. Saville (1985a) *The Economic Case for Deep-Mined*

References

Coal in Scotland: an Interim Report Presented to the National Union of Mineworkers in Scotland, Edinburgh: NUM.

Kerevan, G. and R. Saville (1985b) *The Case for Retaining a European Coal Industry*, Brussels: British Labour Group in the European Parliament.

Kerevan, G. and R. Saville (1987) *Privatising the SSEB: a Report on the Consequences for Scotland*, Edinburgh: Napier College.

Kettle, M. (1985a) 'The National Reporting Centre and the 1984 Miners' Strike', in Fine and Millar (eds) (1985).

Kettle, M. (1985b) '1926 and Now: If the Miners Lose Again', *New Socialist*, February.

Kirby, M. (1972) 'Entrepreneurial Efficiency in the British Coal Industry Between the Wars: a Comment', *Economic History Review*, 25, 4: 655–7.

Kirby, M. (1973a) 'The Control of Competition in the British Coal Mining Industry in the Thirties', *Economic History Review*, 26, 2: 273–84.

Kirby, M. (1973b) 'Government Intervention in Industrial Organization: Coal Mining in the Nineteen Thirties', *Business History*, 15, 2: 160–73.

Kirby, M. (1977) *The British Coalmining Industry 1870–1946: a Political and Economic History*, London: MacMillan.

Kleinwort-Grieveson (1988) *UK Coal – the Role of the Private Sector*, London: Kleinwort Grieveson Securities.

Labour Research (1984) 'Miners' Strike: Economics of Coal', 73, 12, December: 292–5.

Layfield, F. (1986) *Sizewell B Public Inquiry*, 8 volumes, London: HMSO.

Lazonick, W. (1981) 'Factor Costs and the Diffusion of Ring Spinning in Britain prior to World War I', *Quarterly Journal of Economics*, XCVI, 4: 491–516.

Lazonick, W. (1983) 'Industrial Organisation and Technological Change: the Decline of the British Cotton Industry', *Business History Review*, 57, 2, Summer: 195–236.

Letwin, O. (1988a) 'International Experience in the Politics of Privatisation', in Walker (ed.) (1988).

Letwin, O. (1988b) *Privatising the World: a Study of International Privatisation in Theory and Practice*, London: Cassell Educational.

Littlechild, S. (1978) *The Fallacy of the Mixed Economy: an Austrian Critique of Economic Thinking and Policy*, London: IEA.

Littlechild, S. (1979) 'Controlling the Nationalised Industries', Series B Discussion Paper No. 56, Department of Economics, University of Birmingham.

Littlechild, S. (1981) 'The Social Costs of Monopoly Power Revisited', *Economic Journal*, 91, June: 348–63.

Littlechild, S. (1983) *Regulation of British Telecommunications' Profitability*, London: Department of Industry.

Littlechild, S. 61986) *Economic Regulation of Privatised Water Authorities* HMSO: Department of Environment, reproduced in part with some further reflection in *Oxford Review of Economic Policy*, 4, 2: 40–67.

Lloyd, C. (1985) 'A National Riot Police: Britain's "Third Force"?', in Fine and Millar (eds) (1985).

References

Lucas, A. (1937) *Industrial Reconstruction and the Control of Competition*, London: Longmans.
McCloskey, D. (1971) 'International Differences in Productivity: Coal and Steel in America and Britain before World War I', in McCloskey (ed.) (1971).
McCloskey, D. (ed.) (1971) *Essays in a Mature Economy, Britain after 1840*, London: Methuen.
McDonald, G. and H. Gospel (1973) 'The Mond–Turner Talks 1927–33: a Study in Industrial Cooperation', *Historical Journal*, 16, 4: 807–30.
McEwen, J. (1977) *Who Owns Scotland?*, Edinburgh: EUSPB.
McGowan, F. (1987) 'Electricity Privatisation: Does Efficiency Matter?', *Electrical Review*, 220, 23, September: 30–31.
MacGregor, I. (1986) *The Enemies Within: the Story of the Miners' Strike 1984/5*, London: Collins.
McIlroy, J. (1985) ' "The Law Struck Dumb?" – Labour Law and the Miners' Strike', in Fine and Millar (eds) (1985).
McKendrick, N. and R. Outhwaite (eds) (1986) *Business Life and Public Policy: Essays in Honour of D.C. Coleman*, Cambridge: Cambridge University Press.
McKinsey (1987) *Strengthening Competitiveness in UK Electronics*, for NEDO, London: McKinsey and Co.
McLachlan, S. (1983) *The National Freight Buyout*, London: MacMillan.
Maddala, G. (1977) *Econometrics*, New York: McGraw-Hill.
Maddocks, I. (1983) *Civil Exploitation of Defence Technology: Report to the Electronics EDC*, London: NEDO.
Marshall, A. (1959) *Principles of Economics*, 8th edition, London: MacMillan.
Marx, K. (1965) *Capital*, London: Lawrence and Wishart.
Marx, K. (1969) *Theories of Surplus Value*, Part II, London: Lawrence and Wishart.
Matthews, R. and J. Sargent (eds) (1983) *Contemporary Problems of Economic Policy: Essays from the CLARE Group*, London: Methuen.
Mayer, C. (1985) 'Recent Developments in Industrial Economics and their Policy Implications', *Oxford Review of Economic Policy*, 1, 3: 1–24.
Mayer, C. and S. Meadowcraft (1985) 'Selling Public Assets: Techniques and Financial Implications', *Fiscal Studies*, 6, 4: 42–56, reproduced in Kay, Mayer, and Thompson (eds) (1987).
Miliband, R. *et al.* (eds) (1986) *Socialist Register 1985/6*, London: Merlin.
Miliband, R. *et al.* (eds) (1987) *Socialist Register 1987*, London: Merlin Press.
Millward, R. (1982) 'The Comparative Performance of Public and Private Enterprise', in Roll (ed.) (1982).
Minford, P. and P. Kung (1984) *The Costs and Benefits of Pit Closures*, Watford: Public Service Research Centre.
Ministry of Reconstruction (1918) *Report of the Coal Conservation Sub-Committee of the Ministry of Reconstruction on Electric Power in Great Britain*, Cd. 9804, London: HMSO.
Mishan, E. (1968) 'What is Producers Surplus?', *American Economic Review*, 58: 1269–82.

Molyneux, R. and D. Thompson (1987) 'Nationalised Industry Performance: Still Third-Rate?', *Fiscal Studies*, 8, 1: 48–82.

Monopolies and Mergers Commission (1980/81) *Central Electricity Generating Board*, HC 315, London: HMSO.

Monopolies and Mergers Commission (1983) *National Coal Board: A Report on the Efficiency and Costs in the Development, Production and Supply of Coal by the NCB*, Cmnd 8920, London: HMSO.

Monopolies and Mergers Commission (1989) *British Coal Corporation: a Report on the Investment Programme*, Cm 550, London: HMSO.

Moore, J. (1986) *Privatisation in the United Kingdom*, London: Aims of Industry.

National Coal Board (1983) *Collieries and Other NCB Activities*, Statistics Department, London: NCB.

National Coal Board (1984) *List of Collieries*, Statistics Department, London: NCB.

National Coal Board (1985) 'New Strategy for Coal', London: Coal Industry Joint Policy Advisory Committee.

National Council for Civil Liberties (1984) *Civil Liberties and the Miners' Dispute*, First Report of the Independent Inquiry, London: NCCL.

NEDC (1986) *Energy Conservation: Furthering the Debate*, 2nd Report of the Steering Committee on Energy Conservation, London: NEDC.

NEDC (1988) *Performance and Competitive Success: Government Information Technology Policies in Competing Countries*, Electronics Industry Sector Group, London: NEDO.

Nef, J. (1932) *The Rise of the British Coal Industry*, I and II, London: Routledge.

Neumann, A. (1934) *Economic Organisation of the British Coal Industry*, London: Routledge.

Newberry, D. (1986) 'Energy Policy Issues After Privatisation', *Centre for Economic Policy Research Discussion Paper*, 109, London: CEPR.

Nicholas, S. (1982) 'Total Factor Productivity Growth and the Revision of post-1870 British Economic History', *Economic History Review*, 35, 1: 85–98.

O'Donnell, K. (1985) 'Brought to Account: the NCB and the Case for Coal', *Capital and Class*, 26: 105–23.

O'Donnell, K. (1988) 'Pit Closures in the British Coal Industry: a Comparison of the 1960s and 1980s', *International Review of Applied Economics*, 2, 1: 62–77.

Orchard, J. (1922) 'The Rent of Mineral Lands', *Quarterly Journal of Economics*, 36: 290–318.

Pallister, D. et al. (1987) *South Africa Inc.: The Oppenheimer Empire*, London: Simon and Schuster.

Paull, C. (1968) 'Mechanisation in British and American Bituminous Coal Mines. 1890–1939', Unpublished M. Phil. Thesis, University of London.

Pearson, G. (1983) *Hooligans: A History of Respectable Fears*, London: MacMillan.

Percy-Smith, J. and P. Hillyard (1985) 'Miners in the Arm of the Law: a Statistical Analysis', in Scraton (ed.) (1985).

References

Postan, M. and J. Hatcher (1985) 'Population and Class Relations in Feudal Society', in Aston and Philpin (eds) (1985).
Power in Europe (1988) 'UK Electricity Privatisation: 1. A Charter for Change. . . . 2. An Invitation to a Big Fix', 7 December.
Price-Waterhouse (1987) *Privatisation: the Facts*, London.
Privatisation (1986) *Privatisation: a Fact File*, Walton-on-Thames: Tertiary Publications.
Pryke, R. (1982) 'The Comparative Performance of Public and Private Enterprise', *Fiscal Studies*, 3, 2: 68–81.
Redwood, J. (1988) 'Merchant Banks and Privatisation', in Butler (ed.) (1988).
Rees, R. (1984) *Public Enterprise Economics*, London: Weidenfeld and Nicolson, 2nd edition.
Reid Report (1945) *Report of the Technical Advisory Committee on Coal Mining*, Cmd. 6610, London: HMSO.
Report of the Royal Commission on Mining Subsidence (1926/7) two Reports, Cmd. 2570 and 2899, London: HMSO.
Reports of the Royal Commission on Mining Royalties (1890–93), five Reports, with Minutes of Evidence and Appendices, C. 6195, 6531, 6529, 6979, and 6980, London: HMSO.
Rhodes, E. (1945) 'Output, Labour and Machines in the Coalmining Industry in Great Britain', *Economica*, 12: 101–10.
Ricardo, D. (1971) *Principles of Political Economy*, London: Pelican.
Robinson, C. (1988) 'A Liberalised Coal Industry', in Johnson (ed.) (1988), reproduced from *Lloyds Bank Review*, April, 1987.
Robinson, C. and A. Sykes (1987) *Privatise Coal: Achieving International Competitiveness*, Policy Study no 85, London: Centre for Policy Studies.
Robson, W. (ed.) (1937) *Public Enterprise: Developments in Social Ownership and Control in Great Britain*, London: New Fabian Research Bureau.
Roll, E. (ed.) (1982) *The Mixed Economy*, London: MacMillan.
Rooke, N. (1937) *The Ownership of Coal Royalties*, St. Albans: Fisher, Knight and Co.
Rose, H. (1985) 'Securing Social Security', *New Socialist*, March.
Royal Commission on Coal Supplies (1903–5) *Reports of the Royal Commission Appointed to Inquire into the Subject of the Coal Resources of the United Kingdom*, various Cd. numbers, London: HMSO.
Rudig, W. (1986) 'Energy Conservation and Electricity Utilities: a Comparative Analysis of Organisational Obstacles to CHP/DH', *Energy Policy*, 14, 5, October: 104–16.
Samuel Report (1926) *Report of the Royal Commission on the Coal Industry*, Cmd 2600, three volumes, London: HMSO.
Samuel, R. (1985) 'A Managerial Power Cut', *New Socialist*, April.
Samuel, R. et al. (1986) *The Enemy Within*, London: Routledge and Kegan Paul.
Sankey Report (1919) *Reports of the Royal Commission on the Coal Industry, with Minutes of Evidence*, Cmnd 359–361, London: HMSO.
Saville, J. (1986) 'An Open Conspiracy: Conservative Politics and the Miners' Strike 1984/5', in Miliband et al. (eds) (1986).

References

Schmalansee, R. (1979) *The Control of Natural Monopolies*, Lexington: Lexington Books.

Schmalansee, R. (1988) 'Industrial Economics: an Overview', *Economic Journal*, 98, September: 643-81.

Schwartz, M. and R. Reynolds (1983) 'Contestable Markets: An Uprising in the Theory of Industry Structure: Comment', *American Economic Review*, 73: 488-90.

Scott Report (1919) *Ministry of Reconstruction. Third Report of the Acquisition and Valuation of Land for Public Purposes of Rights and Powers in Connection with Mines and Minerals*, Cmd 361, London: HMSO.

Scraton, P. (1985) 'The State versus the People: an Introduction', in Scraton (ed.) (1985).

Scraton, P. (ed.) (1985) *The State versus the People: Lessons from the Coal Dispute*, special issue of *Journal of Law and Society*, 12, 3, Winter.

Sedgemore, B. (1980) *The Secret Constitution*, London: Hodder and Stoughton.

Sen, A. (1973) 'Behaviour and the Concept of Preference', *Economica*, 40, August: 241-59.

Sen, A. (1983) 'Poor, Relatively Speaking', *Oxford Economic Papers*, 35: 153-69.

Sen, A. (1987a) 'Food and Freedom', Text of the Third Sir John Crawford Memorial Lecture, Washington DC: World Bank.

Sen, A. (1987b) *On Ethics and Economics*, Oxford: Blackwell.

Sharkey, W. (1982) *The Theory of Natural Monopoly*, Cambridge: Cambridge University Press.

Shepherd, W. (1984)' "Contestability" vs Competition', *American Economic Review*, 74, 4: 572-87.

Sinfield, A. and N. Fraser (1983) 'The Real Cost of Unemployment', BBC North East, mimeo.

Smith, R. (1961) *Sea-Coal for London*, London: Longmans.

Smith, R. (1987) *Unemployment and Health: a Disaster and a Challenge*, Oxford: Oxford University Press.

Smout, T. (1964) 'Scottish Landowners and Economic Growth, 1650-1850', *Scottish Journal of Political Economy*, 11, November: 218-34.

Sorley, W. (1889) 'Mining Royalties and their Effects on the Iron and Coal Trades', *Journal of the Royal Statistical Society*, 52: 60-98.

Steele, H. (1967) 'Natural Resource Taxation: Resource Allocations and Distribution Implications', in Gaffney (ed.) (1967).

Stelzer, I. (1988) 'Britain's Newest Import: America's Regulatory Experience', *Oxford Review of Economic Policy*, 4, 2: 69-79.

Sunday Times Insight (1985) *Strike*, London: Coronet Books.

Supple, B. (1986) 'Ideology or Pragmatism? The Nationalisation of Coal, 1916-46', in McKendrick and Outhwaite (eds) (1986).

Supple, B. (1987) *The History of the British Coal Industry, Volume 4, 1913-46: The Political Economy of Decline*, Oxford: Clarendon.

Sweet, C. (1985) 'Why Coal is under Attack: Nuclear Powers in the Energy Establishment', in Beynon (ed.) (1985).

References

Sweezy, P. (1933) 'A Note on Relative Shares', *Review of Economic Studies*, 1: 67–8.

Sweezy, P. (1934) 'Professor Pigou's Theory of Unemployment', *Journal of Political Economy*, 42, December: 800–11.

Sweezy, P. (1937) 'On the Definition of Monopoly', *Quarterly Journal of Economics*, 51, February: 362–3.

Sweezy, P. (1938a) *Monopoly and Competition in the English Coal Trade: 1550–1850*, Westport: Greenwood, 1972 reprint.

Sweezy, P. (1938b) 'Wages Policy', *American Economic Review*, Papers and Proceedings, 28, March: 156–7.

Sweezy, P. (1939) 'Demand under Conditions of Oligopoly', *Journal of Political Economy*, 47, April: 263–70.

Sweezy, P. (1942a) *The Theory of Capitalist Development*, New York: Monthly Review Press.

Sweezy, P. (1942b) 'The Illusion of the "Managerial Revolution" ', *Science and Society*, 1, 1: 1–23.

Sweezy, P. (1943) 'Professor Schumpeter's Theory of Innovation', *Review of Economics and Statistics*, 25, February: 93–6.

Sweezy, P. (1949) 'Fabian Political Economy', *Journal of Political Economy*, 57, June: 242–48.

Sweezy, P. (1978) 'Critique', in Hilton (ed.) (1978).

Taussig, F. (1939) *Principles of Economics*, II, New York: MacMillan.

Taylor, A. (1968) 'The Coal Industry', in Aldcroft (ed.) (1968).

Taylor, B. (1984) 'The State of the Art', in Grieve Smith (1984).

Thomas, I. (1937) 'The Coal Mines Reorganisation Commission', in Robson (ed.) (1937).

Thomas, S. (1988) 'Markets for Power Plant Life Extension', European Study Conferences Limited, UK Electricity Opportunities in the 1990s, London.

Thompson, D. (1987) 'De-Regulation and the Advantage of Incumbency', *European Economic Review*, 31: 368–74.

Thompson, F. (1966) 'The Social Distribution of Landed Property in England since the Sixteenth Century', *Economic History Review*, XIX, 3: 505–17.

TUC (1986) *Bargaining in Privatised Companies*, London.

TUC (1987) *Privatisation and Top Pay*, London.

Turner, J. (ed.) (1984) *Businessmen and Politics*, London: Heinemann.

Veljanovski, C. (1987) *Selling the State: Privatisation in Britain*, London: Weidenfeld and Nicolson.

Vickers, J. (1985) 'Strategic Competition among the Few – Some Recent Developments in the Economics of Industry', *Oxford Review of Economic Policy*, 1, 3: 39–62.

Vickers, J. and G. Yarrow (1985) *Privatisation and the Natural Monopolies*, London: Public Policy Centre.

Vickers, J. and G. Yarrow (1988) *Privatisation: an Economic Analysis*, Cambridge: the MIT Press.

Wade, E. (1985) 'The Miners and the Media: Themes of Newspaper Reporting', in Scraton (ed.) (1985).

References

Walker, M. (ed.) (1988) *Privatisation: Tactics and Techniques*, Vancouver: Fraser Institute.

Walker, W. (1986) 'Information Technology and Energy Supply'', *Energy Policy*, 14, 6, December: 466-88.

Ward, J. (1971) 'Landowners and Mining', in Ward and Wilson (eds) (1971).

Ward, J. and R. Wilson (eds) (1971) *Land and Industry: The Landed Estate and the Industrial Revolution*, Newton Abbot: David and Charles.

Weeks, J. (1981) *Capital and Exploitation*, London: Edward Arnold.

Weitzman, M. (1983) 'Contestable Markets: an Uprising in the Theory of Industry Structure: Comment', *American Economic Review*, 73: 486-7.

Welsh Campaign for Civil Liberties (1985) *Striking Back*, Cardiff: WCCPL and NUM (South Wales Area).

Wessel, R. (1967) 'A Note on Economic Rent', *American Economic Review*, 57: 1221-26.

Whitfield, D. (1985) 'Coal: a Privatisation Postponed?, *Capital and Class*, 25, Spring: 5-14.

Williams, R. (1985) 'Mining the Meaning', *New Socialist*, March.

Williamson, O. (1985) *The Economic Institutions of Capitalism: Firms, Markets and Relational Contracting*, London: MacMillan.

Wiltshire, K. (1987) *Privatisation: the British Experience*, Melbourne: Longman Cheshire.

Winterton, J. (1985) 'Computerized Coal: New Technology in the Mines', in Beynon (ed.) (1985).

Winterton, J. (1988) 'The Effect of New Technologies on the Productivity and Costs of the British Coal Industry', *Working Environment Research Group*, Report no. 12, University of Bradford.

Winterton, J. and R. Winterton (1989) *Coal, Crisis and Conflict: the 1984-85 Miners' Strike in Yorkshire*, Manchester: Manchester University Press.

Yarrow, G. (1986) 'Privatisation in Theory and Practice', *Economic Policy*, 2: 324-377.

Yarrow, G. (1988a) 'The Price of Nuclear Power', *Economic Policy*, 6, April: 81-132.

Yarrow, G. (1988b) *Some Economic Issues Surrounding the Proposed Privatisation of Electricity Generation and Transmission*, London: Prima Europe.

Index

abolition of royalties 40, 54-5
accounting 157-8, 198
Acquisition and Valuation of Land Committee 41, 77
Adam Smith Institute 116
Adeney, M. 158
Alsthom 148
amalgamations 22-5, 192
   economies of scale 83, 87, 95
   impeded by cartels 19, 20, 21, 31-2
Americas, the 39
Anglo-American 182
area companies 183-4, 200
artificial monopoly 123, 124
Ascher, K. 112
Ashworth, W. 65
Aspinwall, 52
Association of Chief Police Officers (ACPO) 162
Aston, T. 3
Australia 39, 153
Austria-Hungary 38, 39
Austrian School 112, 117-19, 132, 138

Baker, J. 140, 141
Ball, M. 148
bar machinery 93-4, 97
Baran, P. 3, 4, 6, 10, 11, 15
barriers, coal 61, 77
Bates Colliery 196
Baumol, W. 123
Beesley, M. 132
Belgium 39, 142

Benwell CDP 17
Berry, T. 157, 198
Beynon, H. 133, 155
Birkenhead, Lord 19
Blake, N. 162
Blanden, M. 115
Board of Trade
   amalgamations 22, 23, 24
   cartels 21, 26, 28, 29, 30-1
Bollard, A. 145
Boyfield, K. 184
Boyns, T. 88
Braverman, H. 3
Brenner, R. 15
Breusch, T. 106
British Airways 114
British Coal 111
   accounting 198
   advanced technology 166, 180
   debt 153-4
   decline in output 155
   diversification 182
   miners' strike 161, 163
   privatization 179-81, 182, 183-4, 185
   uneconomic pits 157, 198 costs of closure 167, 169
   see also National Coal Board
British Gas 114, 133, 183
British Leyland 129, 155
British Steel 155
British Telecom 114, 119-20, 135, 175, 183
Brown, A. 115
Brown, L. 158

Index

Brunskill, I. 57, 59
Buckland, R. 115
Buxton, N. xii, 87, 95, 103, 192
   productivity increase 83, 84
Byatt, I. 130-1

Cadeby Colliery 167-9, 170, 172, 196
Cairns, R. 195
Callinicos, A. 158
Cambridge critique 73-4, 81
capability 128-9
capital/labour ratios 73
capitalism 15, 173
   Marx and 159-60
   transition to 15-18
   see also monopoly capitalism
Cardowan Colliery 158
Carsberg, B. 137
cartels xi, 8-9, 19-32, 95
   featherbedding 19, 20-2
   power station construction industry 148
   statutory 20-1, 26-31
   voluntary 25-6
Cecchini, P. 149
Celynen South Colliery 152
Central Collieries Commercial Association 25-6
Central Council of Colliery Owners 26, 28, 29
Central Electricity Generating Board (CEGB) 148, 184
   coal pricing 153, 154, 179, 198
   nuclear power 141, 142, 171
Centre for Policy Studies (CPS) 116
chain machinery 96, 97, 193
   fast adoption of 100, 101, 107
   productivity 93-4
Christian, L. 162
Church, R. 64, 65, 190
Clare Group 121
Clarke, R. 123
cliometrics xi-xii, 81-2
Coal Commission 56
coal cutting, mechanization of 93-4, 96-7
Coal Mines Act (1930) 20-1, 22, 25, 26

Coal Mines Reorganization Commission (CMRC) 20, 62-3, 89, 190
coal preparation 88-9
collective action 103
collieries
   closures 139, 161, 167-9, 196, 198
   mechanization 97-8, 100, 101-2, 107-8
   size and productivity 84-5, 91, 192
   uneconomic 150-8, 170-1, 184
combination 6-8
   see also cartels; monopoly
Combined Heat and Power (CHP) 146-7
Combustion Engineering 148
Commissioners of Crown Lands 56
Company of Hostmen 8, 17, 18
compensation 54, 56-8, 64, 68, 190
competition 73
   Littlechild and 119-20
   monopoly and 8-14
   synthesis and 122-3, 126
   yardstick 136
conduct regulation 136
Conisborough 167, 168
conservation of energy 145-6
Consolidated Gold Field (Consgold) 182
constant returns to scale 72-3
contestability 123, 194
Cook, A.J. 164
co-ordination, inter-district 30
Cortonwood Colliery 161, 199
costs of closures 151-2, 167-9
courts 162, 164
Cowey, 52
Crafts, N. 81, 131
Crick, M. 158
criminalization 163-4, 199
Cromar, P. 187
Cumberbatch, G. 150, 199

Dasgupta, P. 49
Davies, G. 198
Dearne Valley 167, 168
decline

219

Britain's economy 176
  coal productivity 36, 177-8
demand curves 13
denationalization *see* privatization
Denmark 145
Department of Energy 185
destructibility 46
Dick, B. 115
diffusion of mechanization 98-108, 193
Dilnot, A. 128
Dintenfass, M. 64, 88-9, 165
direct provision 128, 195
disc machinery 93-4, 97
Dobb, M. 3, 5-6, 15, 17
Dornbusch, R. 121
Dounreay nuclear power station 143
drainage systems 78

Ecclesiastical Commissioners for England 56
economies of scale 83-91, 95
  estimating 84-7
  regressions for 91-3
Economist 160
efficiency, energy 145-6
Eldon Barry, E. 52
electricity
  British Coal's diversification 182
  privatization 114, 138, 139, 143-50, 196-7
  thermal cost 153
Employee Share Ownership Schemes 127
Employment Act (1980) 160
Employment Protection (Consolidation) Act (1978) 167
energy conservation/efficiency 145-6
Energy Economist 142
Energy Efficiency Office 146
Energy Select Committee 139
  energy saving 146
  nuclear power 142, 143
  privatization of coal 179-80, 181, 185
  privatization of electricity 144-5, 197
entrepreneurship
  failure of interwar industry 89
  mechanization 95, 97-8, 102-3
  ownership 131-2
  price-cutting 13
environment 118
equilibrium 11-12, 13-14
  see also general equilibrium theory; partial equilibrium
Esso 182
European Economic Community (EEC) 168, 170-2
excess capacity 83, 86
Executive Boards 26, 30
exhaustible resources, economics of 50
extraction costs 47-8, 49

Fare, R. 139
featherbedding cartels 19, 20-2
feudalism 16
finance, pooling of 88
Financial Times 161
Fine, B. 10, 22, 173
  economic decline 176
  exhaustible resources 50
  nationalized industries 130, 175, 178
  NUM 180
Fitzwilliam, Earl 64
Five Counties Scheme 25-6
Flinn, M. 6-7, 8
Flux, A. 47
Forrest, R. 112
France 115, 142, 145
  royalties 38, 39, 56
Fraser, N. 168
Friedman, M. 111

gas 153, 177
GEC 148, 149
General Electric 148
general equilibrium theory 5, 11-12
  royalty/rent debate 43-5, 47, 48, 49
General Strike (1926) 140, 164-6
Germany 21, 38, 39, 142, 143, 145
Gibbon, P. 158-9, 200
Glyn, A. 151-2, 154, 155, 167, 198
Goodman, G. 158

Index

Gospel, H. 165
government
    contribution to economy 74-5
    costs of closure to: central 167-8, 169; local 168-9
Government Communications Headquarters (GCHQ) 161
Gowers, E. 20, 63, 190-1
Grand Allies 8
Gray, L. 48, 189
Greasley, D. 102, 193
Greater London Council (GLC) 195
Green, A. 158
Grimethorpe Colliery 143
group equilibrium 13

Habbakuk thesis 81
Hagen, K. 123
Hardie, K. 54, 190
Harris, L. 176
Haslam, 52
Hausmann, W. 8, 187
Hay, D. 121
Hayek, F.A. von 117
Heal, G. 49
Helm, D. 122, 126, 130, 132
    area companies 200
    regulation and competition 133
    state and the economy 194-5
    welfare 128
Henley, A. 31
Hillman, M. 145
Hilton, R. 3, 15, 16
Hobson, J.A. 189
Hoffman-La Roche 118
Holland 145
Holmes, A. 138, 149
Hope, C. 148
Hostmen, Company of 8, 17, 18
housing 68
human capital theory 189
Humberside Ports 139
Hyman, H. 115

imported coal 170-1
Ince, M. 149
income support 128
incumbents 131-2

India 39
industrial policy 176
information technology 175
innovation 14
    see also mechanization
Institute of Economic Affairs (IEA) 116
Institute of Fiscal Studies 121
investment 103, 178, 179
    uneconomic pits 156-7, 198
issue prices 115-16
Italy 38

Japan 143
Jevons, H. 199-200
Jevons, W.S. 43, 199
    decline 36, 177
    royalty/rent debate 36, 44, 188
Johnson, W. 192
joint production 125
Jones, C. 160
Jones, P. 197
Jones, R. 191

Kahn, E. 146
Kay, J. 126, 130, 132, 133, 136, 194
    industrial policy 121, 129
    state intervention 122
Kerevan, G. 153, 158, 169
Kettle, M. 162, 164
Keynes, J.M. 14
Keynesianism 166
Kirby, M. 20, 88, 192
    cartels xi, 95; amalgamations 22-5, *passim*, 31-2;
    Coal Mines Act 21, 26
Kung, P. 198

labour demand 86, 92
Labour Party 113, 164, 166
Labour Research 153
laissez faire 117, 118, 119-20
landed property
    British system and mining 38-9, 75-80, 81
    confrontation between owners and mine owners 61
    miners' attitudes 51, 52

owners running mines 59-60, 80, 190
royalties 56, 65-6, 66-7; Royal Commission 40
large coal 78
Layard, R. 121
Leyfield, F. 141
Lazonick, W. 72, 191, 195
leaseholders 59
Letwin, O. 113, 114, 115, 116
Levene, P. 193
licensing authorities 185-6
Limitation of the Vend 8
Littlechild, S. 117-20, 132, 193
Lloyd, J. 158
local government 168-9
Lucas, A. 29
Luxemburg 38

MacGregor, I. 155, 158, 159, 200
machine index 93-4
Maddala, G. 192
Mahabi, D. 195
management
  British Coal 179, 200
  nationalized industries 127
management buy-outs (MBOs) 127, 193
marginalism 43
market failure 121-2
market forces 11-12
marketing, privatization and 184
Marshall, A. 43, 47, 189
Marx, K. 10, 13
  capitalism 17, 159-60
  rent theory 42, 51, 67-8
Marxism 3, 5-6
Matthews, R. 121
Mayer, C. 115
McCloskey, D. xii, 71, 80-1
  land 78
  productivity 72-3, 75-6
McDonald, G. 165
McGowan, F. 146
McGuiness, T. 123
McIlroy, J. 162
Meadowcroft, S. 115
mechanization 95-108, 192
  cartels 31
  diffusion: Britain as a whole 98-100, 193; districts 101-2; econometric estimates 104-8
  economies of scale 83, 84-5, 88, 91
  reorganization and 166
  royalties 79-80
  Scotland 60
media 164, 199
Meek, R. 3
merchant capital 17
Mercury 133, 183
Metcalf, D. 198
Miliband, R. 112
mine owners
  cartels 20, 26, 32, 103
  control of coal trade 17-18
  Lord Birkenhead's criticism 19
  royalties 41, 59-61, 66, 79-80
Mineral Owners Association of Great Britain (MOAGB) *also* Mineral Association of Great Britain (MAGB) 64, 89
Mineral Owners' Joint Committee 64
miners' attitudes to royalties 41, 51-5, 190
Miners Federation of Great Britain (MFGB) 52, 53, 190
miners' strike (1984/5) xii-xiii, 140, 158-64, 198, 199
  General Strike and 164-6
  uneconomic pits 150, 151
Mines (Working Facilities and Support) Act (1923) 77, 78
Minford, P. 198
Mining Industry Act (1926) 22
Ministry of Reconstruction 41, 61-2
Minorco 182
MINOS facility 183
Molyneux, R. 130
Mond-Turner talks 164-5, 166
monopolies x, 6-8, 118-19
  competition and 8-14
  natural 123-5, 130
  see also monopoly capitalism
Monopolies and Mergers Commission 155-6, 179
monopoly capitalism x-xi, 3, 6, 10-14, 15

Index

Monthly Review School 3
Moore, J. 115-16
motivation 127
multinational corporations 165-6, 182
Murfin, A. 10, 22, 50
Murie, A. 112

National Coal Board (NCB) 68, 111, 179, 183
  see also British Coal
National Economic Development Council (NEDC) 146
National Graphical Association (NGA) 161
National Power 144
National Reporting Centre (NRC) 162
National Trust 118
National Union of Mineworkers (NUM) 150, 151, 161, 163, 180
nationalized industries 138, 174-5
  performance 129-30, 195
  synthesis and 132-3
nationalization 130, 165
  coal industry 65, 68, 125
  land 52
  royalties *see* royalties
natural monopoly 123-5, 130
Nef, J. 7, 17, 18, 56
Neumann, A. 21
New Domesday Survey 190
New Economic History 71, 80-2
New Right 112, 113-14
Newberry, D. 198
Northard, J.H. 179
Novak, T. 160
nuclear power 141-3, 144, 196, 197

O'Donnell, K. 130, 152, 173, 175, 178
oil 153, 177
Orchard, J. 47
output regulation 8-9, 21, 26-9, 30-1
Overseas Development Agency 111
Oxford Review of Economic Policy 121

Pagan, A. 106

Parkinson, C. 111, 181
partial equilibrium
  natural monopoly 124, 194
  royalty/rent debate 43, 44-5, 46, 47-8, 49
Paull, C. 88, 89
percussive machinery 93-4, 97
performance
  collieries 157
  privatization and 129-30, 134, 195
Philpin, C. 3
Pickard, B. 52-3
picketing 161-3
piece rates 78
Pigou, A.C. 12
Pirie, M. 113
Plan for Coal 151, 160
planning 132-3, 135
Plessey 149
police 162-3, 164, 199
Portugal 38
Powell Duffryn Colliery Company 22
Power in Europe 147
Power Gen 144
power station construction 148-9
Price-Waterhouse 115
prices
  coal 170-1; CEGB and 153, 154, 179, 198; nuclear power 142; regulation 8-9, 29, 30, 135; royalties and 37, 47
  electricity 147
  issue 115-16
privatization xii, 111, 173-7
  British Coal 181-6
  electricity 114, 138, 139, 143-50, 196-7
  motives 111-12
  synthesis and 124-5, 126-34
  Thatcher Government and 111, 113-16, 139-40, 175
  wages and 124-5
producers *see* mine owners
production
  joint 125
  ownership of royalties and 54-5
  relations of 16-17, 187

223

Index

production functions 72, 73-4, 81, 85-7, 191
productivity
　decline 36, 177
　increases 155-7
　machine 94
　mechanization and 83, 84-5, 86, 88, 192
profitability 11-12, 64
property rights 118, 126-7, 159-60
Public Record Office 63
public service ethic 127
pumping of water 62, 78

quota transfer 29

rationality 102, 103
rationalization 21-2, 88, 192
redistribution 128-9
redundancies *see* unemployment
Redundant Mineworkers Payments Scheme (RMPS) 167-8
Redwood, J. 113, 115, 197
Rees, R. 121, 132
refurbishment 147
regulation 119-20, 134-7, 173-4, 194, 196
　output 8-9, 21, 26-9, 30-1
　prices 8-9, 29, 30, 135
　privatization and 119-20, 122, 126, 133
Reid Report 19, 79
rent theory xi, 44, 189
　Marx 42, 67-8
　Ricardo 45-6
　see also royalties; royalty/rent debate
research and development (R&D) 142-3, 149-50
reserves 75-6
Ricardo, D. 42, 45-6, 177, 188
Ridley, N. 160
Robinson, C. 184
Rockefeller, J.D. 149, 187
Royal Commission on Coal Supplies 61
Royal Commission on Mining Royalties 36-42, 46, 60, 188
　miners' attitudes 51, 52-3, 54

Royal Commission on Mining Subsidence 77
royalties xi
　distribution of ownership 55-60, 76-7
　fertility and 79-80
　nationalization of 35, 40, 41-2, 62-5, 67, 68; miners' attitudes 41, 51-5, 190
　privatization and 184-5
　problems of system 61-2, 65-6, 78-9, 89-91
　see also Royal Commission on Mining Royalties
royalty/rent debate 45-50
Ryan International 182

safety, nuclear power and 196
Samuel, R. 158
Samuel Report 41, 61, 62, 63-4, 166
Sankey Report 41, 56, 64
Sargent, J. 121
Saville, J. 151
Saville, R. 153, 158, 169
scale, economies of *see* economies of scale
scarcity value 47-8, 49
Scargill, A. 151, 164, 199
Schmalansee, R. 122
Schumpeter, J. 14
Scotland 80, 84, 161, 190
　ownership of royalties 57-60, 76
　price regulation 26, 29
　Southern Electricity Board 179
Scott, L. 77
Scott Report 41, 77
Scottish Coal Marketing Scheme 26
Scraton, P. 199
seams
　depth and thickness 76
　width of cut 79-80
Selzer, S. 158
Sen, A. 127, 128-9
Silbertson, A. 126
Simons, M. 158
Sinfield, A. 168

Index

Sizewell nuclear power station 141, 142, 171
Smillie, R. 54, 190
Smith, A. 12
Smout, T. 59-60
Social Security Acts 160
socialism 132-3
Sorley, W. 47
South Africa 153, 171, 182-3, 191
South Yorkshire 88
Spain 38, 142
stagnation 11, 36
Standard Oil 187
state intervention 121-2, 194-5
statutory marketing schemes 20-1, 26-31
Steele, H. 189
Stock Exchange 120
structural regulation 136
subsidence 77-8
subsidy 128, 129, 171-2
Sunday Times Insight 158
Supple, B. 89-91, 177
Sweezy, P. x, 3-6, 8
   monopoly and competition 8-15
   transition to capitalism 15-18
Sykes, A. 184
synthesis 112, 120-34, 138, 144

'tacksmen' 59
Taussig, F. 48
Taylor, A. 73
technical progress 87, 92-3, 93-4
   see also mechanization
technology, new and privatization 175, 180, 183
Thatcher, M. 112, 113, 164, 181
   'enemy within' 163-4
   uneconomic pits 150
Thatcher Government
   privatization 111, 113-16, 139-40, 175
   unions 160, 162-3
thermal cost 153
Thomas, I. 20
Thomas, S. 147
Thompson, D. 121, 122, 126, 130, 132, 133

Thompson, F. 59
Three Mile Island 142
timber 46
total factor productivity 71, 72-5, 81, 130-1
trade unions 88, 164
   miners' attitudes to royalties 41, 51-5, 190
   ownership of industry 126
   Thatcher Government and 160, 162-3
   see also Miners Federation of Great Britain; National Union of Mineworkers; Union of Democratic Mineworkers
Trades Union Congress (TUC) 52
transition debate 15-18
transportation 6-7, 37

Ulph, D. 132
uneconomic pits 150-8, 170-1, 184
unemployment 139, 175-6
   costs of 151-2, 167-8, 198
   privatization and 175-6, 184
   wages and 12
unification of royalties 62-3
Union of Democratic Mineworkers (UDM) 180, 182
United Kingdom Atomic Energy Authority (UKAEA) 143
United States of America 75-6, 134, 142, 143

Veljanovski, C. 113, 126-7, 127-8
Vickers, J. 121, 129, 132, 136, 174
voluntarily formed cartels 25-6

Wade, E. 164
wages 53, 74
   privatization and 124-5
   unemployment and 12
Wales 25, 29, 80, 88, 161
Walker, M. 113, 114, 115
Walker, P. 139, 181
Walras, L. 43
Ward, J. 59
water, drainage of 62, 78

Index

water authorities 120
Weeks, J. 15, 122
welfare 128-9
Welsh Associated Collieries 23
Wessel, R. 44, 189
Wiltshire, K. 112
Winterton, J. 158, 183
Winterton, R. 158

Woods, 52

yardstick competition 136
Yarrow, G. 121, 123, 126, 132, 174
Yorkshire, South 88
Young, 55